Advances in Spatial Science

Springer
Berlin
Heidelberg
New York
Barcelona
Hong Kong
London
Milan
Paris
Singapore
Tokyo

Titles in the Series

C. S. Bertuglia, M. M. Fischer and *G. Preto* (Eds.)
Technological Change, Economic Development and Space
XVI, 354 pages. 1995. ISBN 3-540-59288-1 (out of print)

H. Coccossis and *P. Nijkamp* (Eds.)
Overcoming Isolation
VIII, 272 pages. 1995. ISBN 3-540-59423-X

L. Anselin and *R. J. G. M. Florax* (Eds.)
New Directions in Spatial Econometrics
XIX, 420 pages. 1995. ISBN 3-540-60020-5 (out of print)

H. Eskelinen and *F. Snickars* (Eds.)
Competitive European Peripheries
VIII, 271 pages. 1995. ISBN 3-540-60211-9

J. C. J. M. van den Bergh, P. Nijkamp and *P. Rietveld* (Eds.)
Recent Advances in Spatial Equilibrium Modelling
VIII, 392 pages. 1996. ISBN 3-540-60708-0

P. Nijkamp, G. Pepping and *D. Banister*
Telematics and Transport Behaviour
XII, 227 pages. 1996. ISBN 3-540-60919-9

D. F. Batten and *C. Karlsson* (Eds.)
Infrastructure and the Complexity of Economic Development
VIII, 298 pages. 1996. ISBN 3-540-61333-1

T. Puu
Mathematical Location and Land Use Theory
IX, 294 pages. 1997. ISBN 3-540-61819-8

Y. Leung
Intelligent Spatial Decision Support Systems
XV, 470 pages. 1997. ISBN 3-540-62518-6

C. S. Bertuglia, S. Lombardo and *P. Nijkamp* (Eds.)
Innovative Behaviour in Space and Time
X, 437 pages. 1997. ISBN 3-540-62542-9

A. Nagurney and *S. Siokos*
Financial Networks
XVI, 492 pages. 1997. ISBN 3-540-63116-X

M. M. Fischer and *A. Getis* (Eds.)
Recent Developments in Spatial Analysis
X, 434 pages. 1997. ISBN 3-540-63180-1

R. H. M. Emmerink
Information and Pricing in Road Transportation
XVI, 294 pages. 1998. ISBN 3-540-64088-6

P. Rietveld and *F. Bruinsma*
Is Transport Infrastructure Effective?
XIV, 384 pages. 1998. ISBN 3-540-64542-X

P. McCann
The Economics of Industrial Location
XII, 228 pages. 1998. ISBN 3-540-64586-1

L. Lundqvist, L.-G. Mattsson and *T. J. Kim* (Eds.)
Network Infrastructure and the Urban Environment
IX, 414 pages. 1998. ISBN 3-540-64585-3

R. Capello, P. Nijkamp and *G. Pepping*
Sustainable Cities and Energy Policies
XI, 282 pages. 1999. ISBN 3-540-64805-4

M. M. Fischer and *P. Nijkamp* (Eds.)
Spatial Dynamics of European Integration
XII, 367 pages. 1999. ISBN 3-540-65817-3

M. M. Fischer, L. Suarez-Villa and *M. Steiner* (Eds.)
Innovation, Networks and Localities
XI, 336 pages. 1999. ISBN 3-540-65853-X

J. Stillwell, S. Geertman and *S. Openshaw* (Eds.)
Geographical Information and Planning
X, 454 pages. 1999. ISBN 3-540-65902-1

G. J. D. Hewings, M. Sonis, M. Madden and *Y. Kimura* (Eds.)
Understanding and Interpreting Economic Structure
X, 365 pages. 1999. ISBN 3-540-66045-3

A. Reggiani (Ed.)
Spatial Economic Science
XII, 457 pages. 2000. ISBN 3-540-67493-4

D. G. Janelle and *D. C. Hodge* (Eds.)
Information, Place, and Cyberspace
XII, 381 pages. 2000. ISBN 3-540-67492-6

P. W. J. Batey and *P. Friedrich* (Eds.)
Regional Competition
VIII, 290 pages. 2000. ISBN 3-540-67548-5

B. Johansson, Ch. Karlsson and *R. R. Stough* (Eds.)
Theories of Endogenous Regional Growth
IX, 428 pages. 2001. ISBN 3-540-67988-X

G. Clarke and *M. Madden* (Eds.)
Regional Science in Business
VIII, 363 pages. 2001. ISBN 3-540-41780-X

Manfred M. Fischer · Yee Leung
Editors

GeoComputational Modelling

Techniques and Applications

With 72 Figures
and 44 Tables

Springer

Professor Dr. Manfred M. Fischer
Vienna University of Economics
and Business Administration
Department of Economic
Geography & Geoinformatics
Rossauer Lände 23/1
1090 Vienna
Austria

Professor Dr. Yee Leung
The Chinese University of Hong Kong
Department of Geography
Shatin, Hong Kong
China

ISBN 978-3-642-07549-0

Library of Congress Cataloging-in-Publication Data applied for
Die Deutsche Bibliothek – CIP-Einheitsaufnahme
Geocomputational: Techniques and Applications; with 44 Tables / Manfred M. Fischer; Yee Leung (ed.). – Berlin; Heidelberg; New York; Barcelona; Hong Kong; London; Milan; Paris; Singapore; Tokyo: Springer, 2001
(Advances in Spatial Science)

Springer-Verlag Berlin Heidelberg New York
a member of BertelsmannSpringer Science+Business Media GmbH
http://www.springer.de

Printed in Germany

Hardcover-Design: Erich Kirchner, Heidelberg

42/2202-5 4 3 2 1 0 – Printed on acid-free paper

Dedicated to our friend and colleague
Stan Openshaw
for his continuing efforts
in the field of spatial analysis

Preface

GeoComputation is a new research paradigm which, although still in its early stages, has the potential to radically change research practice in spatial sciences. It is expected to gather speed and momentum in the first decades of the 21st century. The principal driving forces behind this paradigm are four-fold: *first*, the increasing complexity of spatio-temporal systems, whose analysis requires methods able to deal with non-linearity, uncertainty, discontinuity, self-organization and continual adaptation; *second*, the need to find new ways of handling and using the increasingly large amounts of geographical data about the world, much of which is spatially addressed; *third*, the increasing availability of computational intelligence [CI] technologies that are readily applicable to many areas of spatial science, and have the potential of suggesting better solutions to old problems as well as entirely new developments; and *fourth*, developments in high performance computing that are stimulating the adoption of a computational paradigm for problem solving, data analysis and modelling.

This volume illustrates some of the recent developments in the field of geocomputational modelling and aims to disseminate recent research into graduate classrooms. The idea originated from the Conference of the Croucher Advanced Study Institute [ASI] on Neural and Evolutionary Computations for Spatial Data Analysis, Hongkong, 12–17 June 2000. This book is dedicated to our colleague and friend Stan Openshaw – the hidden driving force behind this novel research paradigm – who missed the meeting due to illness.

We are grateful to all those who have contributed to the present volume. Without their willingness to participate, this endeavour could never have been realized. In addition, we should like to thank Ingrid Divis and Thomas Seyffertitz for their capable assistance in coordinating the various stages of preparation of the manuscripts. We are also grateful for the assistance provided by Angela Spence, whose care and attention to the linguistic editing and indexing have considerably enhanced the quality of the work presented here. Finally, we wish to express our gratitude for support from our home institutions, and the generous financial backing provided by the Croucher Foundation.

Vienna, March 2001

Manfred M. Fischer
Vienna University of Economics and Business Administration

Yee Leung
The Chinese University of Hongkong

Contents

1 GeoComputational Modelling – Techniques and Applications: Prologue

Manfred M. Fischer and Yee Leung***
* Department of Economic Geography & Geoinformatics
Vienna University of Economics and Business Administration
** Department of Geography, The Chinese University of Hongkong

1.1 Introduction

GeoComputation may be a research paradigm still in the making, but it has the potential to dramatically change current research practice in the spatial sciences. Different people, however, have different views of this research paradigm. For some it is synonymous with geographical information systems [GIS]. But GeoComputation is not GIS. There does exist a relationship with GIS, defined as geographical information science, but GeoComputation has other relationships too that are just as important, for example, with computer science, pattern recognition and statistics, neurocomputing and computational intelligence technologies.

From our point of view, GeoComputation captures a large-scale computationally-intensive scientific paradigm that embraces – and here we fully agree with Openshaw and Abrahart (2000) – three interrelated components:

- first, *geographical data* without any constraint on what form it takes or its relationship to GIS,
- second, modern *computational techniques* that aim to compute a solution to a problem that involves some form of geographical information and includes mathematical, statistical and/or soft techniques, or a combination of these;
- third, *high level computing hardware* that spans the hardware spectrum from Pentium PCs over workstations to massively parallel supercomputers with teraflop speeds and terabyte memories.

The driving forces are four-fold:
- the increasing complexity of our spatio-temporal systems, whose analysis requires novel methods for modelling non-linearity, uncertainty, discontinuity, self-organization and continual adaptation,

- the need to create new ways of utilizing and handling the increasingly large amounts of spatial information from the GIS and RS [remote sensing] data revolutions,
- the availability of attractive computational [CI] technologies which provide the modelling tools, and
- the advent of high performance computers, especially those associated with the emerging era of parallel supercomputing.

GeoComputation is a paradigm that brings together researchers and practitioners from a wide variety of fields. The major related fields include uncertainty analysis involving statistics, probability and fuzzy set theory, pattern recognition, artificial and computational intelligence, neural and evolutionary computation and geographical information systems. Let us begin by briefly clarifying the role of each field and how they fit together naturally when unified under the goals and applications of GeoComputation. We should stress that no attempt is made at being comprehensive in any sense of the word (Openshaw et al. 2000).

In **uncertainty analysis,** statistics plays an important role primarily in data selection and sampling, data mining, and the evaluation of geocomputational results. Historically, most work in statistics and its spatial variants has focused on evaluation of data fitting and on hypothesis testing. These are clearly relevant to evaluating the results of GeoComputation to filter the good from the bad, as well as within the search for patterns. On the front end of GeoComputation, statistics offers techniques for detecting outliers, smoothing data if necessary, and estimating noise parameters. But, until now, research has dealt primarily with relatively small data sets and addressed relatively small sample problems.

Most of the statistical work has focused on linear models, additive Gaussian noise models, parameter estimation and parametric techniques for a fairly restricted class of models. Search for pattern and regularities in large databases has received little attention, as efforts have concentrated, whenever possible, on closed-form analytical solutions. Issues connected with the interfaces to geo-databases and dealing with massive data sets have only recently begun to be investigated.

Though statistics deals with uncertainty due to randomness, it is inappropriate to analyze the uncertainty due to fuzziness/imprecision. Fuzzy sets in general, and fuzzy logic in particular, provide a framework within which the imprecision of knowledge and data can be formally examined. Its effectiveness has been demonstrated in spatial classification, especially in the presence of imprecise boundaries, the representation of spatial knowledge and inference, knowledge acquisition, and spatial data analysis. The prevalence of fuzziness in spatial knowledge and data also makes fuzzy sets a good candidate for integration with other GeoComputational methodologies resulting in studies such as fuzzy neuro analysis and fuzzy pattern recognition.

In **pattern recognition**, work has historically focused on practical techniques characterized by an adequate mix of rigour and formalism. The most prominent techniques fall under the category of clustering and classification. Pattern recognition contributions are distinguished from statistics by the utilization of computational algorithms, the more sophisticated data structures, and of course the interest in identifying patterns, both parametric and non-parametric. Significant work is being undertaken in areas such as dimensionality reduction, transformation, classification and clustering [including regionalization].

Techniques and methods originating from **artificial intelligence** [AI] have focused primarily on dealing with data at the symbolic [categorical] level, with little emphasis on continuous variables. AI techniques for reasoning provide a powerful alternative to classical density estimation in statistics. These techniques permit prior knowledge about the domain and data to be included in a relatively simple framework. Other areas of AI too, especially knowledge acquisition and representation, and search are relevant to the various steps of a geocomputational process including data mining, data pre-processing and data transformation.

Computational Intelligence [CI] refers to the lowest-level forms of intelligence stemming from the ability to process numerical [low level] data, without explicitly using knowledge in an AI sense. Computational intelligence tolerates imcision and uncertainty in large-scale real world problems in order to achieve tractability, robustness, computational adaptivity, low cost, real-time speed approaching human-like turnaround and error rates which approximate human performance.

Neurocomputing is a field concerned with information processing systems that autonomously develop operational capabilities in adaptive response to an information environment. The primary information structures of interest in neurocomputing are computational neural networks which comprise densely interconnected adaptive processing elements. Several features distinguish neurocomputing from the field of artificial intelligence in the strict sense. *First*, information processing is inherently parallel. Large-scale parallelism provides a way of significantly increasing the speed of information processing. *Second*, knowledge is encoded not in symbolic structures, but in patterns reflecting the numerical strength of connections between the processing elements of the system [i.e. connectionist type of knowledge representation]. Their adaptive nature makes computational neural networks – in combination with a wide variety of learning techniques – very appealing for GeoComputation, especially in application domains where one has little or incomplete understanding of the problem to be solved, but where training data are readily available. These networks are 'neural' in the sense that they may have been inspired by neuroscience but not necessarily because they are faithful models of biological neural phenomena. In fact, the majority of the networks are more closely related to traditional mathematical and/or statistical models, such as non-parametric pattern classifiers, clustering algorithms, statistical regression models, and non-linear filters than they are to neurobiological models.

Evolutionary computation is based on the collective learning process of a population of individuals, where each individual represents a search point [trial] in the space of potential solutions to a given problem. The three main lines of current evolutionary computational research are *genetic algorithms*, *evolution strategies*, and *evolutionary programming*. Each of these strategies initializes the population of individuals randomly and evolves them towards better regions of the search space by means of a stochastic process of selection, mutation and recombination, as appropriate. The three strategies differ in terms of their specific representation, mutation operations and selection procedures. While genetic algorithms emphasize chromosomal operators based on observed genetic mechanisms [e.g. crossover and bit mutation], evolution strategies and evolutionary programming stress the behavioural link between parents and offspring, rather than the genetic link. A standard evolutionary programming algorithm for optimizing computational neural networks operates as follows. The chosen representation for a solution follows from the problem at hand. In real-valued optimization problems such as training neural models, the individual is taken as a string of real values. The algorithm randomly initializes the initial population of networks and scores each with respect to a given error or performance function. Random mutation creates offspring from these parents. That is, each component [parameter] is perturbed by a Gaussian random variable with mean zero and a variance proportional to the parent's error.

The relevance of **geographical information systems** [GIS] and **remote sensing** [RS] **systems** is evident. They provide the necessary infrastructure to store, access, and manipulate raw data. Supporting such operations from the geocomputational perspective may well become an emerging research area in both the GIS and RS communities. Classical database techniques for query optimization and new object-oriented databases make the task of searching for patterns in spatial databases much more tenable.

The rapidly emerging field of GeoComputation has grown significantly in recent years. This growth is driven by a mix of daunting practical needs and strong research interests. GIS technology has enabled institutions to collect and store information from a wide range of sources at rates that were, only a few years ago, considered unimaginable. Examples of this phenomenon abound in a wide spectrum of fields, including Census data and RS-data. For example, NASA's Earth Observing System is expected to return data at the rate of several gigabytes per hour by the end of the century.

Why are today's techniques, methods and models for spatial data not adequate for addressing the new needs of analysis? The answer lies in the fact that there exist serious computational and methodological problems attached to performing spatial data modelling in high-dimensional spaces with very large amounts of data. These challenges are central to GeoComputation and need urgent attention. Without strongly emphasizing development and research in GeoComputation, we run the risk of forfeiting the value of most of the GIS and RS data that we collect and store. We would eventually drown in an ocean of massive data sets that are rendered useless because we are unable to distill the essence from the bulk!

To date, the general development of GeoComputation has been inhibited by the considerable effort needed to develop each GeoComputation application. But there are now some significant successes in the field. Accounts of many of these can be found in Special Issues of *Environment and Planning A*; *Computers, Environment and Urban Systems*; *The Journal of Geographical Systems: Geographical Information, Analysis, Theory and Decision* as well as two recently published books (Openshaw and Abrahart 2000, Longley et al. 1998). The present volume is a further attempt to make such efforts better known. However work is just beginning in this challenging and exciting field. Some of the challenges are practical awaiting proper implementation, while others are fundamentally difficult research problems that are at the heart of fields such as spatial statistics, modelling, optimization, and pattern recognition.

The current state-of-the-art in GeoComputation still relies rather heavily on fairly simple approaches and great enthusiasm! But despite the frequent lack of methodological rigour, powerful results are being achieved, and the commitment of the research community is rapidly growing. The intention of this volume is to illustrate some recent developments in GeoComputation and attempt to establish some general principles which will serve as a sound foundation for further research.

1.2 About this Book

The rapid transition to more powerful methods of geocomputation and the new advances in spatial science are underlined by two recent readers (Openshaw and Abrahart 2000, Longley et al. 1998). The opportunities provided by geocomputational modelling are five-fold:

- *first,* to enable us to analyze larger databases and/or obtain better results by processing finer resolution data,
- *second,* to improve the quality of research results by utilizing computationally intensive procedures to reduce the number of assumptions and remove the simplifications imposed by computational constraints that are no longer relevant,
- *third,* to speed up existing computer-bound activities so that more extensive experiments can be undertaken,
- *fourth,* to develop new approaches and techniques based upon the availability of practical and applicable computational intelligence [CI] technologies with evolutionary computation, cellular automata, adaptive fuzzy systems and computational neural networks as the major components,
- *finally,* to offer powerful computational methods and techniques to analyze complex spatial systems embedded with non-linearity and uncertainty.

These CI-technologies provide the basis for developing novel styles of spatial analysis and modelling to meet the processing requirements of current-day spatial data. The raison d'être of CI-based modelling is to exploit the tolerance for imprecision and uncertainty in large-scale spatial problems, with an approach characterized by tractability, robustness, computational adaptability, real-time analysis and low cost. There are two principal areas of CI that are particularly relevant to geocomputational research: evolutionary computation including genetic algorithms, evolution strategies and evolutionary programming; and neural networks, which are also known as neurocomputing. Biologically-inspired evolutionary computation provides a basis for developing new solutions to complex spatial optimization problems as well as building blocks for new kinds of spatial analysis techniques and models. Much of the recent interest in neural network modelling in spatial analysis stems from the growing realization of the limitations of conventional tools and models as vehicles for exploring patterns and relationship in GIS and RS [remote sensing] environments. The hope has been raised that these limitations may be overcome by the judicious use of neural net approaches. Their attractiveness is not simply a question of the high computation rates provided by massive parallelism. It stems essentially from the following features:

- the greater representational flexibility and freedom from the design constraints of linear models,
- the built-in ability to incorporate rather than ignore the special nature of spatial data,
- the greater degree of robustness or fault tolerance to deal with noisy data, missing and fuzzy information,
- the capacity to deal efficiently with very large data sets, thus providing the prospects of obtaining better results due to the ability to process finer resolution data or real-time analysis.

The present volume aims to reflect the evolution of CI-based geocomputational modelling and to further our understanding of the opportunities and challenges offered. The book is divided into two parts. While PART A seeks to establish some general principles and build a solid foundation for further research and development, giving particular attention to three major classes of geocomputational models: neural network models, evolutionary computation and cellular automata, PART B deals with some important applications of geocomputational modelling, particularly to pattern recognition in remote sensing environments and spatial interaction modelling.

PART A: Concepts, Modelling Tools and Key Issues

Among the computer-based models and techniques of specific interest in GeoComputation are those derived from the field of artificial intelligence [AI] and the more recently defined area of computational intelligence [CI]. The distinction between artificial and computational intelligence is important because our semantic descriptions of models and techniques, their properties, and our expectations of their performance should be tempered by the kinds of systems we want, and the ones we can build (Bezdek 1994).

The first chapter of **PART A**, by *Manfred M. Fischer*, provides a systematic introduction to computational neural network [CNN] models. The computational appeal of neural networks for solving some fundamental spatial analysis problems is summarized and a definition given of computational neural network models in mathematical terms. Three components of a computational neural network – the properties of the processing elements, network topology and learning – are discussed and a taxonomy of computational neural networks presented, breaking neural networks down according to the topology and type of interconnections, and the learning paradigm adopted. The author distinguishes four families of CNNs that seem to be particularly attractive for solving real world spatial analysis problems: backpropagation networks, radial basis function networks, supervised and unsupervised Adaptive Resonance Theory [ART] models, and self-organizing feature map networks. He stresses several important challenges which need to be tackled in the years to come. The first is to develop application-domain-specific methodologies relevant for spatial analysis, and the second to gain deeper theoretical insights into the complex relationship between learning [training] and generalization, crucial for the success of real-world applications. A third challenge is the delivery of high-performance computing on neuro-hardware to enable rapid computational neural network prototyping with the ultimate goal of developing application-domain-specific automatic computational neural network systems. This is crucial for making these models another element in the toolbox of geocomputation.

How to learn from examples, the problem which neural networks were designed to solve, is one of the most important research topics in artificial intelligence. One way of formalizing this form of learning is to assume the existence of a function representing the set of examples, which thus enables generalization. This can be called function reconstruction from sparse data [or in mathematical terms, depending on the required precision, approximation or interpolation, respectively]. Within this general framework, the central issues of interest are the representational power of a given network model [in other words, the problem of model selection or network design] and the procedures for obtaining the optimal network parameters [the problem of network training]. Both problems – model selection or the specification of a network topology and network training or learning – are key methodological issues.

In recent years there has been increasing interest in utilizing evolutionary computation for solving these issues. Evolutionary computation provides powerful classes of probabilistic adaptive algorithms based on principles of biological evolution. They operate on a population of candidate solutions [individuals] in parallel. By operating on a population of trial solutions, evolutionary algorithms explore the search space from many different points simultaneously, and reduce the risk of converging to local optima. They include genetic algorithms, evolution strategies and evolutionary programming. Typical algorithms update the population seeking for better solutions in the search space using operations of selection, recombination and mutation, inspired by biological evolution. The environment delivers a quality information [*fitness value*] of the search points, and the selection process favours those individuals of higher fitness to reproduce more often than worse individuals. The recombination mechanism allows the mixing of parental information while passing it to their descendants, and mutation introduces innovation into the population.

Computational neural networks in which evolution is another fundamental form of adaptation may be referred to as evolutionary computational neural networks [ECNNs]. Evolution can be introduced in CNN modelling at three levels: *first*, the evolution of connection weights [model parameters] provides an adaptive search to connection weight training; *second*, the evolution of neural network architectures is used to automatically generate a near optimal architecture for the task at hand; *third*, the evolution of learning rules allows a computational neural network to adapt its learning rule to its environment so that efficient learning can be obtained.

Chapter 3, written by *Xin Yao*, introduces the concept of ECNNs, reviews the current state-of-the-art and indicates possible future research directions. The author describes a novel evolutionary system, called EPNet, for evolving feedforward CNNs, which learn both their weights and architectures simultaneously using a hybrid algorithm. A non-linear ranking scheme and five mutation operations are used in the algorithm. These five mutation operators are applied sequentially and selectively to each individual in a population. Such sequential application encourages the evolution of smaller CNNs with fewer hidden nodes and connections.

Most evolutionary approaches to CNN design and training use an evolutionary algorithm to minimize a given error [performance] function. The best individual in a population gets most attention and is selected from the evolutionary system as the final output. The rest of the population is discarded. The author argues that a population contains more useful information than the best individual is providing. Such information can be used to improve the performance of evolved CNNs. In essence, one can regard a population of CNNs as a CNN ensemble, and combine outputs from different individuals using various techniques, such as voting and averaging. Some experiments are presented to show the advantages of exploiting the population information.

Chapter 4, written by *Yee Leung*, shifts attention to neural and evolutionary computation in the fields of spatial classification and knowledge acquisition. The author provides a very useful review of a basket of relevant neural network models, such as conventional and fuzzy types of backpropagation neural networks, radial basis function neural nets and linear associative memories along with various parameter estimation procedures, including backpropagation combined with the gradient descent learning rule. A neural-based clustering method called scale-space filtering is proposed as an algorithm for cluster identification, cluster validity check and outlier detection. This differs substantially from the conventional statistical approach to clustering. The chapter also introduces two rule-based expert system shells for rule learning, and shows the role genetic algorithms can play in this framework in general, and in training fuzzy classification systems in particular.

The final chapter of PART A, written by *Michael Batty*, moves to geocomputational modelling based on the theory of cellular automata [CA]. This theory defines systems in any number of dimensions in terms of basic units, called cells, within which the elements comprising the system – called states – exist. The basic principle which drives the system through time is based on the notion that cell states change as a function of what is happening to other cells in their *local* neighbourhood. Such neighbourhoods are composed of adjacent cells and as such, do not admit any action-at-a-distance. Systems with global properties emerging from local actions and interactions between cells can thus be simulated using CA. In spatial problems, the cells usually correspond to real geographical spaces, their configuration being regular, often laid out in regular tesselations, such as a grid.

The chapter introduces a framework for understanding the dynamics of urban growth, and in particular the continuing problem of urban sprawl. The models presented are based on the gradual transition from vacant land to established development. The author suggests that the mechanism of transition is analogous to the way an epidemic is generated within a susceptible population. Waves of development are generated from the conversion of available land to new development, followed later by redevelopment due to the aging process. The standard aggregate model is presented in differential equation form, showing how different variants [including logistic, exponential, predator-prey models] can be derived for various urban growth situations. The model is then generalized to a spatial system and the author shows how sprawl can be conceived as a process of both interaction/reaction and diffusion. Operationalizing the model as a cellular automata [CA], implying that diffusion is entirely local, he then illustrates how waves of development and redevelopment characterizing both sprawl and aging of the existing urban stock, can be simulated. Lastly, the chapter shows how the model can be adapted to a real urban situation – using as an example the Ann Arbor area in Eastern Michigan – where it demonstrates how waves of development are absorbed and modified by historical contingencies associated with the pre-existing urban structure.

PART B: Spatial Application Domains

Neural networks, which may be viewed as non-linear extensions of conventional spatial statistical models, are applicable to two major domains: *firstly*, as universal function approximators to areas such as spatial regression, spatial interaction, spatial choice and space-time series analysis; and *secondly*, as pattern recognizers and classifiers to allow the user to shift through the data intelligently, reduce dimensionality, and find patterns of interest in data-rich environments.

There is little doubt that neural pattern classifiers have an important role to play in high dimensional problems of pattern recognition and the classification of massive quantities of data. Spatial pattern recognition in remote sensing is an enormously complex task. It requires the analysis of data sets in spaces of high dimensionality derived from spatial, temporal and sensor measurement space components. Chapter 6, written by *Graeme Wilkinson*, provides evidence that neurocomputational techniques offer a number of important advantages in extracting products from remote sensing data for both supervised and unsupervised classification tasks. The development of neural network techniques in remote sensing is being driven by the increasing sophistication of sensor technology, such as hyperspectral systems, and by the need for increasingly accurate environmental maps. At present, the most promising techniques for building useful spatial pattern recognition systems are those based on backpropagation neural network architectures or on hybrid methodologies integrating computational neural networks, evolutionary algorithms, fuzzy logic and expert system technologies. Hybrid systems provide the key to achieving significantly better performance than individual component technologies, in solving complex pattern recognition problems involving very large databases and tough time constraints.

The next chapter, written by *Sucharita Gopal* and *Manfred M. Fischer*, shifts attention to the Adaptive Resonance Theory of Carpenter and Grossberg (1987a, b), which is closely related to adaptive versions of k-means such as ISODATA. Adaptive resonance theory provides a large family of models and algorithms, but limited analysis has been performed of their properties in real world environments. The authors analyze the capability of the neural pattern recognition system, fuzzy ARTMAP, to generate high quality classifications of urban land cover using remotely sensed images. Fuzzy ARTMAP synthesizes fuzzy logic and Adaptive Resonance Theory [ART] by exploiting the formal similarity between the computations of fuzzy subsets and the dynamics of category choice, search and learning. The contribution describes design features, system dynamics and simulation algorithms of this learning system, which is trained and tested for classification [with eight classes given a priori] of a multispectral image of a Landsat-5 Thematic Mapper scene [270 x 360 pixels] from the city of Vienna on a pixel-by-pixel basis. The performance of the fuzzy ARTMAP is compared with that of an error-based learning system based upon the multi-layer perceptron, and the Gaussian maximum likelihood classifier as conventional statistical benchmark on the same database. Both neural classifiers outperform the conventional classifier in terms of classification accuracy. Fuzzy ARTMAP leads to out-of-sample classification accuracies which are very close to maximum performance, while the multi-layer perceptron – like the conventional classifier – has difficulty in distinguishing between some land use categories.

Chapter 8, by *Manfred M. Fischer*, focuses on neural spatial interaction modelling. Such models are termed 'neural' in the sense that they are based on the computational models inspired by neuroscience. They are closely related to spatial interaction models of the gravity type and, under commonly met conditions, can be understood as a special class of feedforward neural networks with a single hidden layer and sigmoidal transfer functions. This class of networks can provide approximations with an arbitrary precision, i.e. it has a universal approximation property. Building a neural spatial interaction model involves two distinct tasks: determination of the network topology, and estimation of weights [parameters]. These two tasks are of crucial importance for the success of real world applications and thus deserve special attention. In this chapter, the parameter estimation problem is defined as a problem of minimizing the sum-of-square error function, and a brief characterization given of the local search procedures generally used to solve this problem. Differential evolution, an efficient direct search procedure, is suggested as a powerful alternative to current practice.

The aim of model selection is to optimize the complexity of the model in order to achieve the best generalization. The chapter provides insight into this problem by introducing the concept of bias-variance trade-off, in which the generalization error is broken down into the sum of the bias squared plus the variance. A neural spatial interaction model that is too simple will smooth out some of the underlying structure in the data [producing high bias], while a model that has too much flexibility in relation to the particular data set will overfit the data and have a large variance. In either case, the performance of the model on new data will be poor. This highlights the need to optimize the complexity in the model selection process in order to achieve the best generalization of a model. Three principal ways of controlling the complexity of a model and directing model search are discussed: *network pruning*, the use of *penalty terms* and *stopped* or *cross-validation training*. The first two approaches can be viewed as variations of long established statistical techniques corresponding, in the case of pruning, to specification searches and, in the case of penalty terms, to regularization or biased regression. The procedure of cross-validation training seems to be one of the true innovations to come out of neural network research. The model chosen here does not require the training process to converge. Rather, the training process is used to perform a directed search of parameter space to find a model with superior generalization performance. Randomness enters neural spatial interaction modelling in two ways: the splitting of data samples, and in choices about parameter initialization. This leaves one question wide open, namely what variation in generalization performance is obtained as one varies training, validation and test sets? It is shown that the bootstrapping pairs approach, a computationally intensive non-parametric approach to statistical inference, provides a better statistical picture of the forecast variability of the model.

The final chapter of the book, by *Günter Haag*, concentrates on the application of neurocomputing tools to approximation problems in the domain of mobility panel analysis. The author analyzes the applicability of single hidden layer backpropagation networks using data from the German Mobility Panel 1994 to the prediction of two types of variable: the overall time spent on mobility [a continuous variable] and car usage on a specific day [a discrete variable].

Since neural network models perform essentially arbitrary non-linear functional mappings between sets of variables, a single hidden layer neural network can, in principle, be used to map the raw input data directly onto the required final output values. In practice, for all but the simplest problems, such an approach will generally lead to poor results. For most applications it is crucial to first transform the input data into some new representation before training a neural network. Thus, the author pays attention to the issue of data pre-processing. The variables in the example are on very different scales; some are continuous, others discrete. Each variable is transformed to the unit interval. Ordinal variables are transformed directly into the corresponding values of a continuous variable, while a 1-of-N coding is utilized for the category input data. There are various ways to control the complexity of the functions represented by a network [see Chapter 8 in this volume for details]. Here, the author applies connection weight and unit pruning to arrive at appropriate feed forward neural networks which tend to slightly outperform conventional statistical models.

Like most good research, the contributions in this volume raise as many questions as they answer. The interaction between a sound theoretical and methodological foundation and careful data analysis seems to be the key to continued progress in geocomputational modelling in order to meet the new challenges of expanding geocyberspace. It is hoped that the insights provided here will inspire other scholars to further explore the world of GeoComputation.

PART A: Concepts, Modelling Tools and Key Issues

2 Computational Neural Networks – Tools for Spatial Data Analysis

Manfred M. Fischer
Department of Economic Geography & Geoinformatics,
Vienna University of Economics and Business Administration

2.1 Introduction

The proliferation and dissemination of digital spatial databases, coupled with the ever wider use of Geographic Information Systems [GIS] and Remote Sensing [RS] data, is stimulating increasing interest in spatial analysis from outside the spatial sciences. The recognition of the spatial dimension in social science research sometimes yields different and more meaningful results than analysis that ignores it.

Spatial analysis, as it has evolved over the last few decades, basically includes two major fields: spatial data analysis and spatial modelling, though the boundary is rather blurred. Spatial data analysis may be defined as a core technology [i.e. body of methods and techniques] for analyzing events [objects] where the results of analysis depend on the spatial arrangement of the events (see Haining 1994). These may be represented in the form of a point, line or area, in the sense of spatial primitives located in geographical space, to which a set of one or more attributes are attached. Location, topology, spatial arrangement, distance and spatial interaction have become a major focus of attention in activities concerned with detecting patterns in spatial data, and with exploring and modelling the relationships between such patterns.

Empirical studies in the spatial sciences routinely employ data for which locational attributes are an important source of information. Such data characteristically consist of one or more cross-sections of observations for either micro-units such as individuals [households, firms] at specific points in space, or aggregate spatial entities such as census tracts, electoral districts, regions, provinces, or even countries. Observations such as these, for which the absolute location and/or relative positioning [spatial arrangement] is explicitly taken into account, are termed *spatial data*. In the socioeconomic realm points, lines, and areal units are the fundamental entities for representing spatial phenomena. This form of spatial referencing is also a salient feature of GIS and *Spatial Data Infrastructures*. Three broad classes of spatial data can be distinguished:

- *object data* where the objects are either *points* [spatial point patterns or locational data, i.e. point locations at which events of interest have occurred] or *areas* [area or lattice data, defined as discrete variations of attributes over space],
- *field data* [also termed geostatistical or spatially continuous data], which consists of observations associated with a continuous variation over space, or given values at fixed sampling points, and
- *spatial interaction data* [sometimes called link or flow data] consisting of measurements each of which is associated with a link or pair of locations representing points or areas.

Analyzing and modelling spatial data present a series of problems. Solutions to many of them are obvious, others require extraordinary efforts. Data can exercise a considerable power, and misinterpreted they may lead to meaningless results; therein lies the tyranny of data. The validity of quantitative analysis depends on *data quality*. Good data are reliable, contain few or no mistakes, and can be used with confidence. Unfortunately, nearly all spatial data are flawed to some degree. Errors may arise in measuring both the location and attribute properties, but may also be associated with computerized processes responsible for storing, retrieving, and manipulating the data. The solution to the quality problem is to take the necessary steps to avoid faulty data determining research results.

The particular form [i.e. size, shape and configuration] of the spatial aggregates can affect the results of the analysis to a varying, but usually unknown degree (see, e.g., Openshaw and Taylor 1979, Baumann, Fischer and Schubert 1983). This problem has become generally recognized as the *modifiable areal units problem* [MAUP], the term stemming from the fact that areal units are not 'natural' but usually arbitrary constructs. For reasons of confidentiality, social science data [e.g. census data] are not often released for the primary units of observation [individuals], but only for a set of rather arbitrary areal aggregations [enumeration districts or census tracts]. The problem arises whenever area data are analyzed or modelled, and can lead to two effects: one derives from selecting different areal boundaries while holding the overall size and the number of areal units constant [*the zoning effect*]. The other derives from reducing the number but increasing the size of the areal units [*the scale effect*]. There is no analytical solution to the MAUP, but the following type of question have to be considered in constructing an areal system for analysis: What are the basic spatial entities for defining areas? What theory guides the choice of the spatial scale? Should the definition process follow strictly statistical criteria and merge basic spatial entities to form larger areas using some regionalization algorithms (see Wise, Haining and Ma 1996)? These questions pose daunting challenges (Fischer 2001).

Most of the spatial data analysis techniques and methods currently in use were developed in the 1960s and 1970s, i.e. in an era of scarce computing power and small data sets. Their current implementation takes only limited advantage of the data storage and retrieval capabilities of modern computational techniques, and basically ignores both the emerging new era of parallel supercomputing and computational intelligence techniques.

There is no doubt that spatial analysis in general and spatial data analysis in particular is currently entering a period of rapid change. It promises to be a period which presents a unique opportunity for developing novel styles of data analysis which will respond to the need to efficiently and comprehensively explore large databases for patterns and relationships, against a background of data uncertainty and noise, especially when the underlying database is of the order of gigabytes. In this chapter, we hope to provide a systematic introduction to computational neural networks [CNNs] with the aim of helping spatial analysts learn about this exciting new field. The hope is that this contribution will help to demystify and popularize computational neural networks in spatial analysis, and stimulate neural network applications in spatial analysis which are methodologically and technically sound, rather than ad hoc. Some of the material in this chapter is similar to introductions to certain computational neural networking in spatial analysis that have already appeared (Fischer 1995, 1998b). Some parts have also been adapted from Fischer and Abrahart (2000).

The chapter is structured as follows. In Section 2.2 a brief summary is given of the computational appeal of neural networks for solving some fundamental spatial analysis problems. This is followed by a definition of computational neural network models in mathematical terms [Section 2.3]. The next three sections are devoted to a discussion of the three fundamental components of a computational neural network: the properties of the processing elements [Section 2.4], network topology [Section 2.5] and learning [determination of the weights on a connection] in a computational neural network [Section 2.6]. In Section 2.7 a taxonomy of CNNs is presented, breaking neural networks down according to the topology and type of interconnections [network topology] and the learning [training] paradigm adopted. The concluding section pays attention to the question of how neurocomputing techniques differ from conventional statistics.

2.2 Why Computational Neural Networks?

Briefly stated, a CNN model is a parallel, distributed information structure consisting of a set of adaptive processing [computational] elements and a set of unidirectional data connections [a more detailed definition is given below]. These networks are neural in the sense that they have been inspired by neuroscience, but not necessarily because they are faithful models of biological neural or cognitive phenomena. In fact, the networks covered in this contribution are more closely related to traditional mathematical and/or statistical models such as non-parametric pattern classifiers, statistical regression models and clustering algorithms than they are to neurobiological models.

The notion computational neural networks is used to emphasize rather than to ignore the difference between computational and artificial intelligence. Neglect of this difference might lead to confusion, misunderstanding and misuse of neural network models in spatial analysis. Computational intelligence [CI] denotes the lowest-level forms of 'intelligence' which stem from the ability to process numerical [low-level] data without using knowledge in the artificial intelligence [AI] sense. An AI system (exemplified in a spatial analysis context by spatial expert systems: see Smith et al. 1987; Kim, Wiggins and Wright 1990; Webster 1990; Leung 1993) is a CI system where added value comes from incorporating knowledge in form of non-numerical information, rules and constraints that humans process. Thus, the neural networks considered in this chapter, such as feedforward and pattern classifiers and function approximators, are computational rather than AI systems.

Why have neural network based spatial analysis been receiving increasing attention in the last few years? There are a number of reasons. From a computational point of view, they offer four primary attractions which distinguish them qualitatively from the standard information approaches currently in use in spatial analysis.

The strongest appeal of CNNs is their suitability for *machine learning* [i.e. computational adaptivity]. Machine learning in CNNs consists of adjusting connection weights to improve the performance of a CNN. This is a very simple and pleasing formulation of the learning problem. Certainly *speed of computation* is another key attraction. In traditional single processor von Neuman computers, the speed is limited by the propagation delay of the transistors. Computational neural networks, on the other hand, because of their intrinsic massively parallel distributed structure, can perform computations at a much higher rate, especially when these CNNs are implemented on parallel digital computers, or, ultimately, when implemented in customized hardware. This feature makes it possible to use CNNs as tools for real-time applications involving, for example, pattern recognition in GIS and RS data environments. It is clear that the increasing availability of parallel hardware will enhance the attractiveness of CNNs and other basically parallel models in spatial analysis.

Furthermore, because of their non-linear nature, CNNs can perform functional approximation and pattern classification operations, which are beyond optimal linear techniques. CNNs offer a *greater representational flexibility* and *freedom from linear model design.* They are semi- or non-parametric and make weaker assumptions concerning the shape of underlying distributions than conventional statistical models. Other important features include *robust behaviour* with noisy data. Noise refers to the probabilistic introduction of errors into data. This is an important aspect of real-world applications, and CNNs can be especially good at handling noise in a reasonable manner.

CNNs may be applied successfully in a variety of diverse areas in spatial analysis to solve the problems of the following type:

Pattern Classification

The task of pattern classification is to assign an input pattern represented by a feature vector to one of several prespecified classes. Well-known applications include spectral pattern recognition utilizing pixel-by-pixel spectral information to classify pixels [resolution cells] from multispectral imagery to a priori given land cover categories. As the complexity of the data grows [use of more spectral bands of satellite scanners, and finer pixel and grey level resolutions], so too does the need for more powerful pattern classification tools.

Clustering or Categorization

In clustering, also known as unsupervised pattern classification, there are no training data with known class labels. A clustering algorithm explores the similarity between the patterns and places similarity between patterns in a cluster. Well-known clustering applications include data mining, data compression, and exploratory spatial data analysis. Very large scaled pattern classification tasks based on census small area statistics for consumer behaviour discrimination, for example, have proven to be difficult in unconstrained settings for conventional clustering algorithmic approaches, even using very powerful computers (see Openshaw 1995).

Function Approximation

Suppose a set of n labelled training patterns [input-output pairs], $\{(x_1, y_1), (x_2, y_2), ..., (x_n, y_n)\}$ have been generated from an unknown function $\Phi(x)$ [subject to noise]. The task of function approximation is to find an estimate, say Φ^*, of the unknown function Φ. Various spatial analysis problems require function approximation. Prominent examples are spatial regression and spatial interaction modelling.

Prediction/Forecasting

Given a set of n samples $\{y(t_1), y(t_2), ..., y(t_n)\}$ in a time sequence $t_1, t_2, ..., t_n$, the task is to predict the sample $y(t_n+1)$ at some future time t_n+1. Prediction/ forecasting has a significant impact on decision-making in regional development and policy.

Optimization

A wide variety of problems in spatial analysis can be posed as [non-linear] spatial optimization problems. The goal of an optimization algorithm is to find a solution satisfying a set of constraints such that an objective function is maximized or minimized. The Travelling Salesman Problem, an NP-complete problem, and the problem of finding optimal locations are classical examples.

2.3 Definition of a Computational Neural Network

Computational neural networks are fundamentally simple [generally non-linear] adaptive information structures. However, it is helpful to have a general definition first before characterizing the building blocks of these structures in more detail. In mathematical terms a computational neural network model is defined as a directed graph with the following properties:

- a state level u_i is associated with each node i,
- a real-valued weight w_{ij} is associated with each edge ji between two nodes j and i that specifies the strength of this link,
- a real-valued bias θ_i is associated with each node i,
- a [usually non-linear] transfer function φ_i [u_i, w_{ij}, θ_i, $(i{\neq}j)$] is defined for each node i. This determines the state of the node as a function of its bias, the weights of its incoming links, and the states of the nodes j connected to it by these links.

In the standard terminology, the nodes are called *processing elements* [PEs], processing, computational units or neurons. The edges of the network are called connections. They function as unidirectional conduction paths [signal or data flows] that transmit information in a predetermined direction. Each PE can receive any number of incoming connections called *input connections*. This property is referred to as an unlimited fan-in characteristic of the PEs, i.e., the number of incoming connections into a node is not restricted by some upper bound. Each PE can have any number of output connections, but carries an identical PE state level [also called activity level, activation or output signal]. The weights are termed connection parameters and determine the behaviour of the CNN model.

For illustration purposes, a typical CNN architecture is pictured in Fig. 2.1. In the figure the circles denote the processing elements, and the arrows the direction of the signal flow. The single output signal of the PEs branches into copies which are distributed to other PEs or which leave the network altogether. The network input represented by the external world can be viewed as data array $\boldsymbol{x}$, $x = (x_1, ..., x_n) \in \Re^n$ [n-dimensional Euclidean space] and the output of the network as data array $y = (y_i, ..., y_m) \in \Re^m$ [m-dimensional Euclidean space with $m<n$]. When viewed in this fashion, the CNN can be thought of as a function Φ: $\Re^n \rightarrow \Re^m$.

The PEs implement transfer functions [i.e. primitive functions] which are combined to produce Φ, the so-called network function. This observation is the basis for the mechanism to embed CNNs into a programmed information system. Computational neural network models are specified by

- the node characteristics [i.e. properties of the processing elements],
- the network topology [i.e. pattern of connections between the processing elements, also termed network architecture], and
- the methods of determining the weights on the connections [called learning rules or algorithms, machine learning, network training or parameter estimation].

Both design procedures involving the specification of node characteristics and the network topology, and learning rules are the topic of much current research.

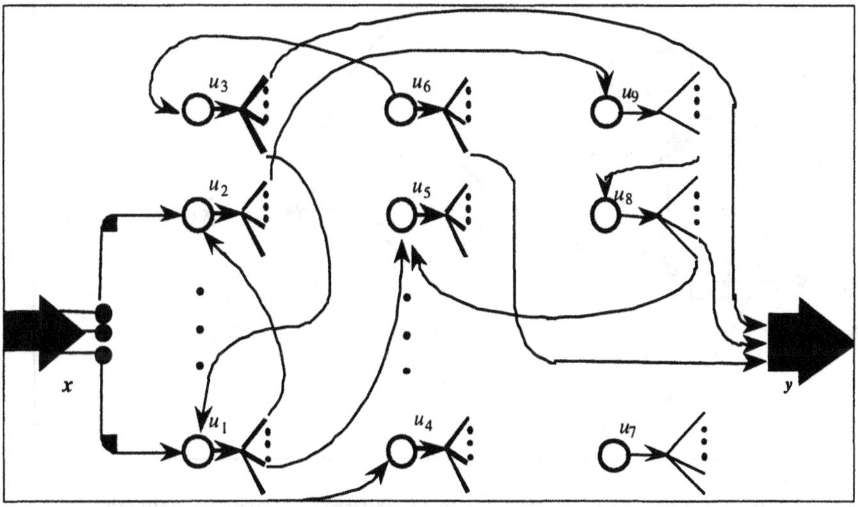

Fig. 2.1 A typical computational neural network architecture

2.4 Properties of the Processing Elements

Processing elements are fundamental to the operation of any CNN. Thus, it is important to have a closer look at the operation of such elements since they form the basis of the CNNs considered in this chapter.

Fig. 2.2 shows internal details of a generic processing unit *i* from an arbitrary computational neural network. The processing unit in this figure occupies a position in its network that is quite general; i.e. this PE accepts inputs from other PEs and send its output [activation] to other PEs. PEs that are neither input nor output units maintain this generality and function as fundamental non-linear computing devices in a CNN.

In order to simplify notation, we use u_i to refer to both the processing unit and the numerical activation [output] of that unit where there is no risk of confusion. Each unit u_i computes a single [numerical] unit output or activation. For example, output u_8 in Fig. 2.1 is both an output of the network as a whole and an input for unit u_5 to be used in the computation of u_i's activation. Inputs and outputs of the processing elements may be discrete, usually taking values {0, 1} or {−1, 0, 1}, or they may be continuous, generally assuming values [0, 1] or [−1, +1]. The CNNs we consider in this chapter are characterized by continuous inputs and outputs.

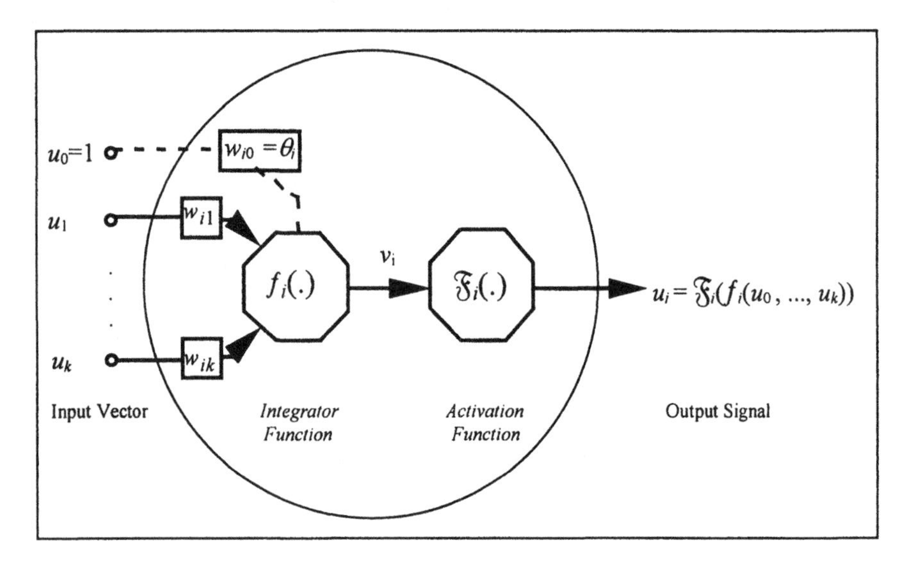

Fig. 2.2 Generic processing element u_i, from an arbitrary computational network

Fig. 2.2 assumes that the processing unit in question, u_i, gets k input signals, $\boldsymbol{u} = \{u_1, ..., u_k\}$, arriving via the incoming connections which impinge on unit u_i. Note that the k units are indexed 1 through k so that $k < i$. The corresponding connection weights are w_{ij} (j = 1, ..., k). It is important to make a note of the manner in which the subscript of the connection weight w_{ij} is written. The first subscript refers to the PE in question and the second subscript refers to the input end of the connection to which the weight refers. The reverse of this notation is also used in the neural network literature. We refer to the weights $\boldsymbol{w}_{i.} = \{w_{i1}, ..., w_{ik}\}$ as the weights of unit u_i. To simplify notation $\boldsymbol{w}$ is used for the vector $\boldsymbol{w}_{i.}$.

Positive weights indicate reinforcement, negative weights represent inhibition. By convention there is a unit u_0, whose output is always +1, connected to the PE in question. The corresponding weight w_{i0} is called bias θ_i of unit i [i.e. the bias is treated as any other weight]. Thus, we may define the (k+1)-by-1 input vector

$$\boldsymbol{u} = [1, u_1, u_2, ..., u_k]^T \tag{2.1}$$

and, correspondingly, we define (k+1)-by-1 weight [also called connection weight, connection parameter or simply parameter] vector

$$w = [\theta, w_{i1}, ..., w_{ik}]^{\mathrm{T}} \tag{2.2}$$

The basic operation of a processing element in computing its activation or output signal u_i involves applying a transfer function φ_i that is composed of two mathematical functions (Fischer 1995): an integrator function f_i and an activation or output function $\mathfrak{F}_i$, i.e.

$$u_i = \varphi_i(\boldsymbol{u}) = \mathfrak{F}_i(f_i(\boldsymbol{u})) \tag{2.3}$$

Typically, the same transfer function is used for all processing units in any particular layer of a computational neural network, although this is not essential.

The integrator function f_i performs the task of integrating the activation of the units directly connected to the processing element in question and the corresponding weights for those connections, thus reducing the k arguments to a single value [called net input to or activity potential of the PE] v_i. Generally, f_i is specified as the inner product of the vectors $\boldsymbol{u}$ and $\boldsymbol{w}$, as follows

$$v_i = f_i(\boldsymbol{u}) = \langle \boldsymbol{u}, \boldsymbol{w} \rangle = \sum_{j=0,1,...,k} w_{ij}\, u_j \tag{2.4}$$

where $\boldsymbol{w}$ has to be predefined or learned during training. In this case the net input to the PE is simply the weighted sum of the separate outputs from each of the k connected units plus a bias term w_{i0}. Because of the weighted process used to compute v_i, we automatically get a degree of tolerance for noise and missing data (see Gallant 1993). The bias term represents the offset from the origin of the k-dimensional Euclidean space $\Re^k$ to the hyperplane normal to $\boldsymbol{w}$ defined by f_i. This arrangement is called a first order processing element because f_i is an affine [linear when w_{i0}=0] function of its input vector $\boldsymbol{u} = (u_1, ..., u_k)^{\mathrm{T}}$. Higher order processing elements arise when more complicated functions are used for specifying f_i. A second order processing unit is, for example, realized if f_i is specified as a quadratic form, say $\boldsymbol{u}^T\, \boldsymbol{w}\, \boldsymbol{u}$, in $\boldsymbol{u}$.

The activation or output function, denoted by $\mathfrak{F}_i\, (f_i\, (.))$, defines the output of the processing unit in terms of the net input v_i at its input. There are various ways of specifying $\mathfrak{F}_i$, but usually it is specified as a non-linear, non-decreasing, bounded [i.e. $\mathfrak{F}_i$ is expected to approach finite maximum values asymptotically] and piece-wise differentiable functional form. For computational efficiency it is desirable that its derivative be easy to compute.

If inputs are continuous, assuming values on [0,1], generally the logistic function chosen is:

$$\mathfrak{F}_i(v_i) = \frac{1}{1+exp(-\beta\, v_i)} \tag{2.5}$$

where β denotes the slope parameter which has to be a priori chosen. In the limit, as β approaches infinity, the logistic function becomes simply a threshold function.

2.5 Network Topologies

In the previous section, we discussed the characteristics of the basic processing element in a CNN. This section focuses on the pattern of connections between the PEs [termed architecture] and the propagation of data. As for the pattern of connection a major distinction can be made between feedforward computational neural networks, and recurrent computational neural networks. Feedforward computational neural networks are networks that do not contain directed cycles. It is often convenient to organize the nodes of a feedforward CNN into layers. We define a $\mathtt{l}$-layer feedforward network as a CNN where processing units are grouped into $\mathtt{l}+1$ subsets [layers] L_0, L_1, ..., $L_\mathtt{l}$ such that if u in a layer L_a is connected to cell u_i in a layer L_b than $a < b$. For a strictly $\mathtt{l}$-layer network, we require additionally that units can be connected only to units in the next layer, i.e. $b = a+1$. All units in a layer L_0 are input units, all trainable units are in layers L_1, ..., $L_\mathtt{l}$, and all units in layer $L_\mathtt{l}$ are output units.

In the simplest form of a layered feedforward CNN, we just have a layer L_0 of input nodes that projects on to a layer L_1 of output computational units. Such a CNN is called a single-layer feedforward network, with the designation 'single-layer' referring to the output layer of PEs. In other words, we do not count the layer of input nodes, because no computation is performed here [i.e. the transfer function being the identity]. Many real world problems, however, need more sophisticated architectures to be adequate, even though interesting theoretical results can be obtained from these simple CNNs.

The second class of feedforward CNNs is distinguished by the presence of one or more hidden layers whose processing elements are correspondingly called hidden elements or units. The function of the hidden units is to intervene between the external input and the network output to extract higher order statistics.

The nodes of the input layer L_0 supply the respective elements of the input vector which constitute the input signals applied to the PEs of the second layer [i.e. the first hidden layer L_1]. The output signals of L_1 are used as the inputs to L_2, and so on for the rest of the network. Characteristically, the PEs in each layer are assumed to have the same transfer function φ and to have the output signals of the preceding layer as their inputs. The set of output signals of the PEs in the output layer L_l constitutes the overall response of the CNN.

The architectural graphs of Fig. 2.3 illustrate the layout of two multilayer feedforward CNNs for the case of a single hidden layer, so-called single hidden layer feedforward networks. For brevity, the networks of Fig. 2.3 are referred to as 8:4:2 networks in that they have eight input nodes, four hidden PEs and two output units.

The CNN of Fig. 2.3(a) is said to be fully connected in the sense that every node of the network is connected to every other node in the adjacent forward layer. If some of the connections are missing from the network, then the network is said to be partially connected. An example of such a CNN is shown in Fig. 2.3(b). Each PE in the hidden layer is connected to a partial set of input nodes in its immediate neighborhood. Such a set of localized nodes feeding a PE is said to constitute the receptive field of the PE. The CNN of Fig. 2.3(b) has the same number of input nodes, hidden units and output units as Fig. 2.3(a), but it has a specialized structure. In real world applications, the specialized structure built into the design of a feedforward CNN reflects prior information about the problem to be analyzed. The procedure for translating a feedforward CNN diagram, that that illustrated in Fig. 2.3, into the corresponding mathematical function Φ follows from a straightforward extension of Section 2.3.

The output of unit i is obtained by transforming the net input, as in Equation (2.4), with a non-linear activation function $\mathfrak{F}_i$ to give

$$u_i = \mathfrak{F}_i \left(\sum_{j=0,1,\ldots,k} w_{ij}\, u_j \right) \tag{2.6}$$

where the sum runs over all inputs and units which send connections to unit i [including a bias parameter]. For a given set of values applied to the inputs of the CNN, successive use of (2.6) allows the activation of all processing elements in the CNN to be evaluated including those of the output units. This process can be regarded as a forward propagation of signals through the network. The result of the whole computation is well-defined and we do not have to deal with the task of synchronizing the processing elements. We just assume that the PEs are visited in a fixed order, each PE re-evaluating and changing its activation before the next one is visited.

A *recurrent* [*feedback*] *computational neural network* distinguishes itself from a feedforward CNN in that it contains cycles [i.e. feedback connections] as illustrated by the architectural graph in Fig. 2.4. The data is not only fed forward but also back from output to input units. This recurrent structure has a profound impact on the learning capability of the CNN and its performance. In contrast to feedforward networks, the computation is not uniquely defined by the interconnection pattern and the temporal dimension must be considered.

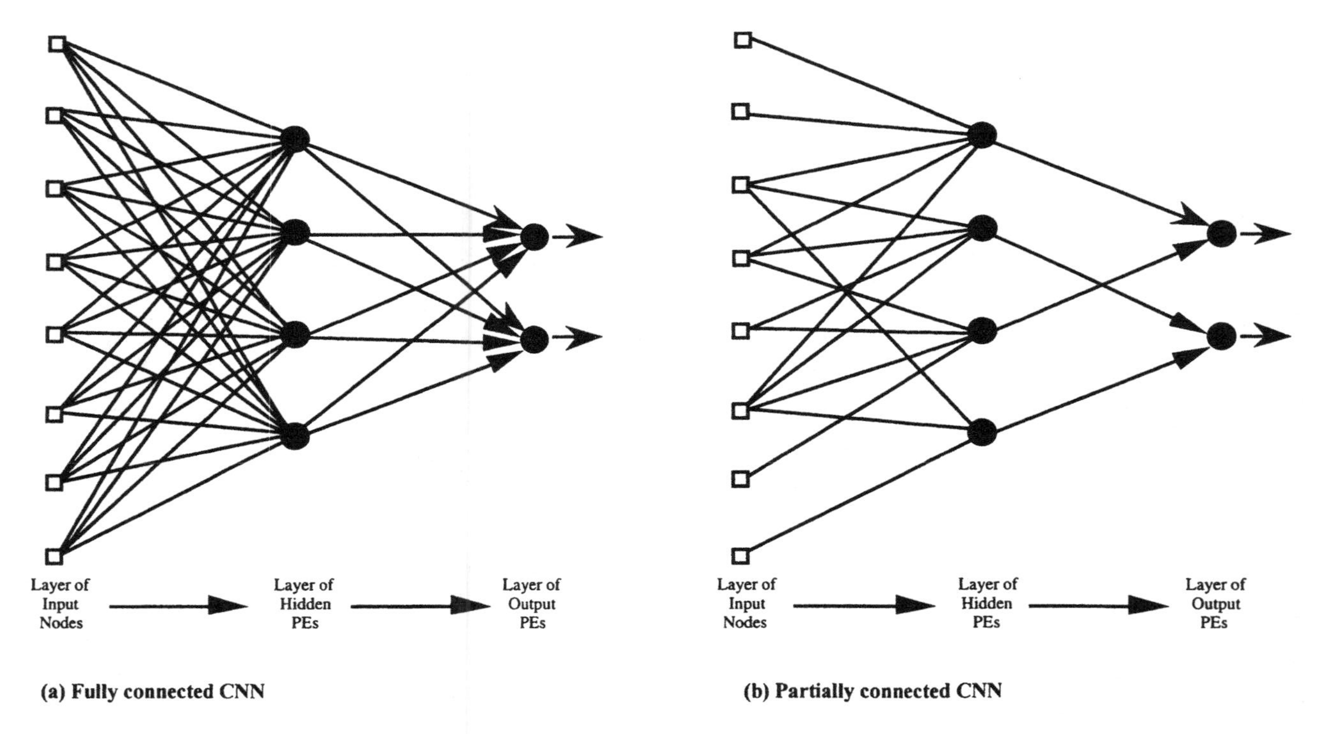

(a) Fully connected CNN

(b) Partially connected CNN

Fig. 2.3 Feedforward computational neural network architectures with one hidden layer: (a) fully connected and (b) partially connected (see Fischer 1998)

When the output of a PE is fed back to the same element, we are dealing with a recursive computation without an explicit halting condition. We must then define what to expect from the CNN. Is the fixed point of the recursive evaluation the desired result or one of the intermediate computations? To solve this problem it is usually assumed that every computation takes a certain amount of time at each node. If the arguments for a PE have been transmitted at time t, its output will be produced at time $t+1$. A recursive computation can be stopped after a certain number of steps and the last computed output takes as the result of the recursive computation.

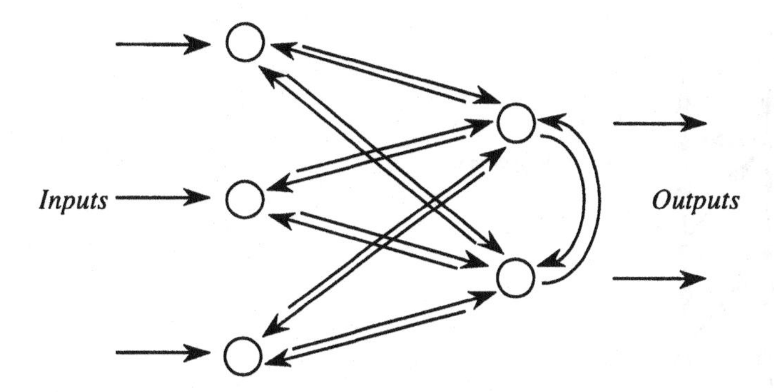

Fig. 2.4 A recurrent [feedback] computational neural network architecture (see Fischer 1998b)

2.6 Learning in a Computational Neural Network

In addition to the characteristics of the processing elements and the network topology, the learning properties are an important distinguishing characteristic of different computational neural networks. In the context of CNNs, the process of learning may be best viewed as a search [typically local step-wise and steepest-gradient-based] in a multidimensional weight space for a solution [i.e. a set of weights] which gradually optimizes a prespecified objective [performance, criterion] function with or without constraints. Learning is normally accomplished through an adaptive procedure, known as learning or training rule, or [machine] learning algorithm.

Each CNN is associated with one or more algorithms for machine learning. The input to these algorithms consists of a finite set of training examples, also called training patterns, $\{E^r\}$, which is an n-vector of values that gives settings for the corresponding units of a CNN model, as illustrated in Fig. 2.5. The j-th component of an example, E_j^r is assigned to input unit u_j ($j = 1, ..., n$).

It is convenient to distinguish two types of learning situations: *first*, supervised learning problems, and *second*, unsupervised learning problems. For *supervised learning problems* [also known as learning with a teacher or associative learning] each example $\boldsymbol{E}^r$ is associated with a correct response $\boldsymbol{C}^r$ [also termed teacher signal] for the CNN's output units. For CNNs with more than one output unit $\boldsymbol{C}^r$ is a vector with components C_j^r (= 1, ..., m). For supervised learning problems the term training example [pattern] is intended to include both the input $\boldsymbol{E}^r$ and the desired response $\boldsymbol{C}^r$.

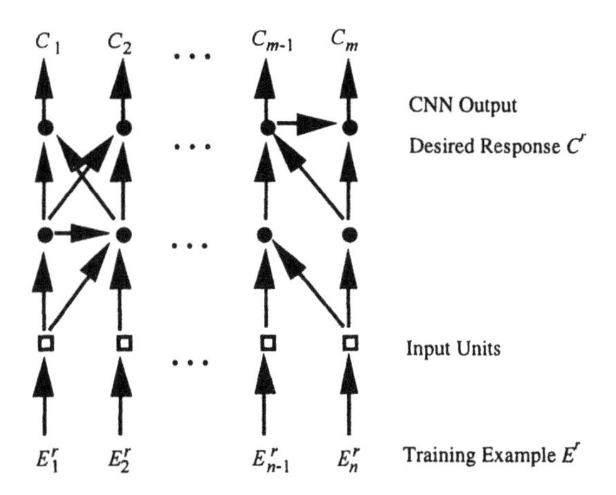

Fig. 2.5 Training example inputs and outputs for a supervised learning model (see Fischer 1998b)

For unsupervised learning problems there are no specified correct network responses available. Unsupervised learning [typically based on some variation of Hebbian and/or competitive learning] generally involves the clustering of, or detection of similarities among, unlabelled patterns of a given training set. The idea here is to optimize some performance [criterion] function defined in terms of the output activity of the PEs in the CNN. The weights and the outputs of the CNN are usually expected to converge to representations that capture the statistical regularities of the training data. Usually the success of unsupervised learning hinges on some appropriately designed CNN that encompasses a task-independent criterion of the quality of representation the CNN is required to learn.

There is a wide variety of learning algorithms, for solving both supervised and unsupervised learning problems, that have been designed for specific network architectures. Many learning algorithms – especially those for supervised learning in feedforward CNNs – have their roots in function-minimization algorithms that can be classified as local or global search [minimization] algorithms. Learning algorithms are termed 'local' if the computations needed to update each weight of a CNN can be performed using information available locally to that weight. This requirement may be motivated by the desire to implement learning algorithms in parallel hardware. Local minimization algorithms, such as those based on gradient-descent, conjugate-gradient and quasi-Newton methods, are fast but usually converge to local minima. In contrast, global minimization algorithms, such as simulated annealing and evolutionary algorithms, have heuristic strategies to help escape from local minima. All these algorithms are weak in either their local or their global search. For example, gradient information useful in local search is not used well in simulated annealing and evolutionary algorithms. In contrast, gradient-descent algorithms with multistarts are weak in global search.

Designing efficient algorithms for CNN learning is a very active area of research. In formulating CNN solutions for [large-scale] real world problems, we seek to minimize the resulting algorithmic complexity and, therefore, the time required for a learning algorithm to estimate a solution from training patterns.

2.7 A Taxonomy of Computational Neural Networks

A taxonomy of four important families of computational neural network models [backpropagation networks, radial basis function networks, supervised and unsupervised ART models, and self-organizing feature map networks] that seem particularly attractive for solving real world spatial analysis problems is presented in Fig. 2.6. This taxonomy is first divided into CNNs with and without directed cycles. Below this, CNNs are divided into those trained with and without supervision. Further differences between CNNs not indicated in Fig. 2.6 refer to the retrieval mode [synchronous versus asynchronous] and the inputs [discrete versus continuous].

Backpropagation computational neural networks have emerged as major workhorses in spatial analysis. They can be used as universal function approximators in areas such as spatial regression, spatial interaction, spatial choice and space-time series analysis, as well as pattern classifiers in data-rich environments [see, for example, Fischer and Gopal 1994a, Fischer et al. 1994, Leung 1997a]. Strictly speaking, backpropagation is a technique which provides a computationally efficient procedure for evaluating the derivatives of the network's performance function with respect to the network parameters, and corresponds to a propagation of errors backwards through the network. This technique was popularized by Rumelhart, Hinton and Williams (1986).

Backpropagation is used primarily with multi-layer feedforward networks [also termed multi-layer perceptrons], so it is convenient to refer to this type of supervised feedforward network as a backpropagation network. Backpropagation CNNs are characterized by the following network, processing element and learning properties:

- *network properties*: multilayer [typically single hidden layer] network with
- *processing element properties*: characteristically continuous inputs and continuous non-linear sigmoid-type PE transfer functions, normally assuming values [0,1] or [−1,+1], where the evaluation of the network proceeds according to the PE ordering, with each PE computing and posting its new activation value before the next PE is examined; the output unit activations are interpreted as the outputs for the entire CNN;
- *learning properties*: the essential ingredient is the backpropagation technique typically, but not necessarily, in combination with a gradient descent based learning algorithm.

The *Radial Basis Function CNNs* are a special class of a single hidden layer feedforward network. Each processing element in the hidden layer utilizes a *radial basis function*, such as a Gaussian kernel, as the transfer function. The argument of the transfer function of each hidden unit computes the Euclidean norm between the input vector and the centre of the unit. The kernel function is centred at the point specified by the weight vector associated with the PE. Both the positions and the widths of these kernels must be learned from training patterns. Each output unit implements characteristically a *linear* combination of these radial basis functions. From the point of view of function approximation, the hidden units provide a set of functions that constitute a basis set for representing input patterns in the space spanned by the hidden units. There is a variety of learning algorithms for the Radial Basis Function CNNs. The basic one utilizes hybrid learning that decouples learning at the hidden layer from learning at the output layer.

This technique estimates kernel positions and kernel widths using a simple unsupervised k-means based clustering algorithm, followed by a supervised least mean square algorithm to determine the connection weights between the hidden and the output layer. Because the output units are typically linear, a non-iterative algorithm can be used. After this initial solution is obtained, a supervised gradient-based algorithm can be used in a further step to refine the connection parameters. It is worth noting that Radial Basis Function Networks require more training data and more hidden units than backpropagation networks to achieve the same level of accuracy, but are able to train much faster – by several orders of magnitude.

ART network models differ from the previous CNNs in that they are recurrent. The data are not only fed forward but also back from the output to the input units. ART networks, moreover, are biologically motivated and were developed as possible models of cognitive phenomena in humans and animals. The basic principles of the underlying theory of these networks, known as adaptive resonance theory [ART], were introduced by Grossberg (1976a, b). They are essentially clustering CNNs with the task to automatically group unlabeled input vectors into several categories [clusters] so that each input is assigned a label corresponding to a unique cluster. The clustering process is driven by a similarity measure, vectors in the same cluster are similar which means that they are close to each other in the input space.

The networks use a simple representation in which each cluster is represented by the weight vector of a prototype unit. If an input vector is close to a prototype, then it is considered a member of the prototype's cluster, and any differences are attributed to unimportant features or to noise. The input and the stored prototype are said to resonate when they are sufficiently similar. There is no set number of clusters, they are created as needed. Any clustering algorithm that does not prespecify the number of clusters must have some parameter that controls cluster granularity. For ART models this parameter is called the vigilance parameter. Regardless of its setting, the ART networks are stable for a finite set of training examples in that final clusters will not change with additional iterations from the original set of training examples. Thus, they show incremental clustering capabilities and can handle an infinite stream of input data. They do not require large memory to store training data because their cluster prototype units contain implicit representation of all the inputs previously encountered. However, ART networks are sensitive to the presentation order of the training examples.

They may yield different clustering on the same data when the presentation order of patterns is varied. Similar effects are also present in incremental versions of classical clustering techniques such as k-means clustering [which is also sensitive to the initial choice of cluster centres].

The basic architecture of an ART network involves three sets of processing elements:

- an input processing field [called the F_1-layer] consisting of two parts: the input portion with input nodes and the interface portion [interconnection],
- the layer of linear units [called the F_2-layer] representing prototype vectors whose outputs are acted as on competitive learning [i.e. the winner is the node with the weight vector which is closest, in a Euclidean distance sense, to the input vector], and
- supplementary units implementing a reset mechanism to control the degree of similarity of patterns placed on the same cluster.

The interface portion of the F_1-layer combines signals from the input portion and the F_2-layer in order to compare the similarity of the input signals to the weight vector for the cluster that has been selected as a candidate for learning. Each unit in the interface portion is connected in the F_2-layer by feedforward and feedback connections. Changes in the activation of units and in weights are governed by coupled differential equations.

ART models come in several varieties, most of which are unsupervised. The simplest are ART 1 designed for clustering training examples $\{\boldsymbol{E}^r\}$ with discrete data, $\boldsymbol{E}_j^r \in \{0,1\}$ [Carpenter and Grossberg 1987a] and ART 2 designed for continuous data $\boldsymbol{E}_j^r \in [0,1]$ [Carpenter and Grossberg 1987b]. A more recent addition to the ART family is ARTMAP, a supervised learning version of ART, consisting of a pair of ART1 modules for binary training and teaching signals linked together via an inter-ART associative memory, called a map field [Carpenter, Grossberg and Reynolds 1991a]. Fuzzy ARTMAP is a direct generalization of ARTMAP for continuous data, achieved by replacing ART 1 in ARTMAP with fuzzy ART. Fuzzy ART synthesizes fuzzy logic and adaptive resonance theory by exploiting the formal similarity between the computations of fuzzy subsethood and the dynamics of prototype choice, search and learning. This approach is appealing because it nicely integrates clustering with supervised learning on the one hand and fuzzy logic and adaptive resonance theory on the other. Chapter 7 in this volume illustrates fuzzy ARTMAP performance in relation to backpropagation and maximum likelihood classifiers in a real world setting of a spectral pattern recognition problem.

Another important class of recurrent networks are self-organizing feature map [SOFM] networks that have been foremost and primarily developed by Kohonen (1982, 1988). One motivation for such topology-preserving maps is the structure of the manmalian brain. The theoretical basis of these networks is rooted in vector quantization theory, the motivation for which is dimensionality reduction. The purpose of SOFM is to map input vectors $\boldsymbol{E}^r \in \Re^n$ onto an array of units [normally one- or two-dimensional] and to perform this transformation adaptively in a topologically ordered fashion such that any topological relationship among input vectors is preserved and represented by the network in terms of a spatial distribution of unit activities. The more these two vectors are related in the input space, the closer one can expect the position of the two units representing these input patterns to be in the array. The idea is to develop a topographic map of the input vectors so that similar input vectors would trigger nearby units. Thus, a global organization of the units is expected to emerge.

The essential characteristics of such networks may be summarized as follows:

- *Network properties*: a two-layer network where the input layer is fully connected to an output layer [also called a *Kohonen layer*] whose units are arranged in a two-dimensional grid [map] and are locally interacting [local interaction means that changes in the behaviour of a unit directly affect the behaviour of its immediate neighbourhood],
- *Processing element properties*: each output unit is characterized by a n-dimensional weight vector; processing elements are linear because each Kohonen unit computes its net input in a linear fashion with respect to its inputs; non-linearities come in when the selection is made as to which unit 'fires'.
- *Learning properties*: unsupervised learning, that consists of adaptively modifying the connection weights of a network of locally interacting units in response to input excitations and in accordance with a competitive learning rule [i.e. weight adjustment of the winner and its neighbours]; the weight adjustment of the neighbouring PEs is instrumental in preserving the order of the input space.

SOFM networks can also be used as a front end for pattern classification or other decision-making processes where the Kohonen layer output can be fed into, for example a hidden layer of a backpropagation network.

It is hoped that the different families of CNN models presented in this chapter will give the reader an appreciation of the diversity and richness of these models. There is no doubt that they provide extremely rich and valuable non-linear mathematical tools for spatial analysts. Their application to real world settings holds the potential for making fundamental advances in empirical understanding across a broad spectrum of application fields in spatial analysis.

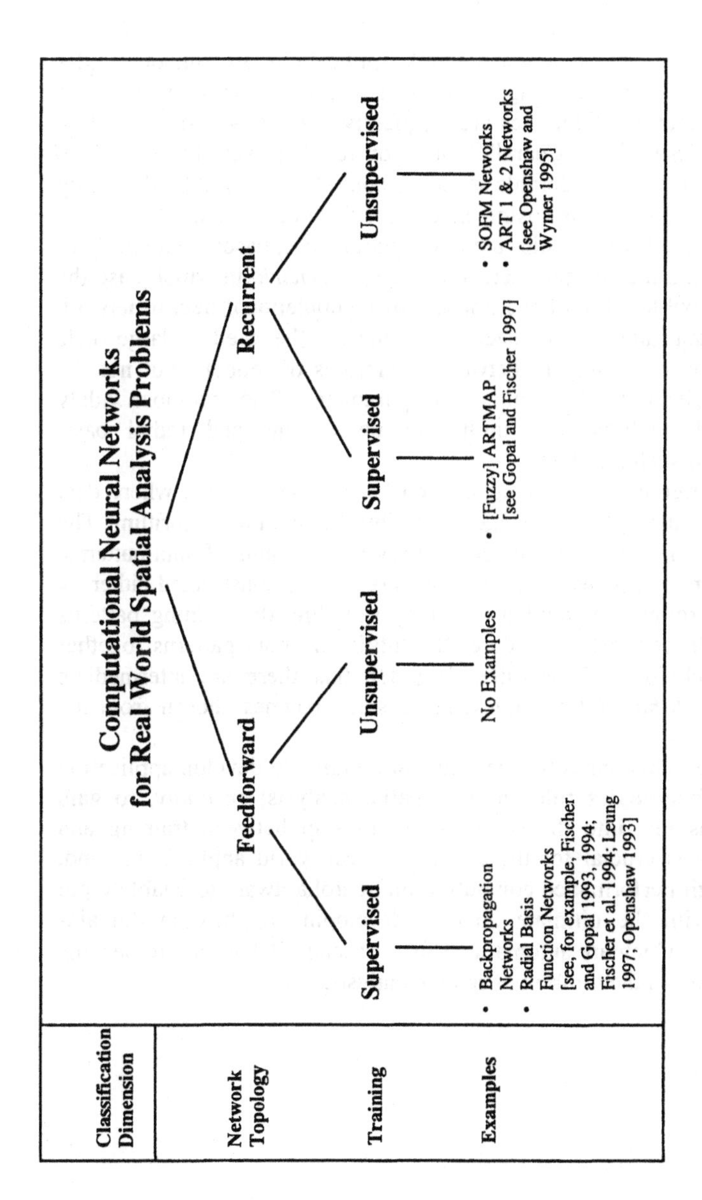

Fig. 2.6 A taxonomy of computational neural networks

2.8 Outlook – How Do Neurocomputing Techniques Differ?

In practice the vast majority of neural network applications are run on single-processor digital computers, although specialist parallel hardware is being developed. Neurocomputing techniques are frequently very slow and a speed-up would be extremely desirable, but parallelisation on real hardware has proved to be a non-trivial problem. A considerable speed-up could be achieved by designing better learning [training] algorithms using experience from other fields.

The traditional methods of statistics are either *parametric*, based on a family of models with a small number of parameters, or *non-parametric* in which case the models are totally flexible. One of the impacts of computational neural network models on spatial data analysis has been to emphasize the need in large-scale practical problems for something in between – families of models, but not the unlimited flexibility given by a large number of parameters. The two most widely used neural network architectures, multilayer perceptrons and radial basis functions, provide two such families.

Another major difference is the emphasis on '*on-line*' procedures where data are not stored, except through the changes made by the learning algorithm. The theory of such algorithms has been studied for very long streams of data patterns, but the practical distinction is less clear, as such streams are constructed either by repeatedly cycling through the training set or by sampling the training patterns [with replacement]. In contrast, procedures that utilize all data patterns together are called '*batch*' techniques. It is often forgotten that there are intermediate positions [called '*epoch-based*' techniques] using small batches chosen from the training set.

The most important challenges for the future are, *firstly*, to develop application domain-specific methodologies relevant for spatial analysis; *secondly*, to gain deeper theoretical insights into the complex relationship between training and generalization, which is crucial for the success of real world applications; and, *thirdly*, to deliver high performance computing on neurohardware to enable rapid CNN prototyping, with the ultimate goal of developing application domain-specific CNN-systems which are automatic. This is crucial if CNNs are to become another regular element in the toolbox of spatial analysts.

3 Evolving Computational Neural Networks Through Evolutionaı y Computation

Xin Yao
School of Computer Science, The University of Birmingham

3.1 Introduction

Computational neural networks [CNNs] have been used widely in many application areas in recent years. Most applications use feedforward CNNs and the backpropagation [BP] training algorithm. There are numerous variants of the classical BP algorithm and other training algorithms, but all these training algorithms assume a fixed CNN architecture. They only train weights in the fixed architecture that includes both connectivity and node transfer functions [see also Chapter 8 in this volume]. The problem of designing a near optimal CNN architecture for an application remains unsolved. This is an important issue, because there is strong biological and engineering evidence to support the contention that the function, i.e. the information processing capability of an CNN, is determined by its architecture.

This chapter is mainly concerned with connectivity and will use the terms architecture and connectivity interchangeably unless stated otherwise. There have been many attempts at designing CNN architectures, especially connectivity, automatically using various constructive and pruning algorithms (Fahlman and Lebiere 1990, Nadal 1989, Burgess 1994, Setiono and Hui 1995, Reed 1993). Roughly speaking, a constructive algorithm starts with a minimal network [i.e. a network with a minimal number of hidden layers, nodes, and connections] and adds new layers, nodes, and connections if necessary during training, while a pruning algorithm does the opposite, i.e. deletes unnecessary layers, nodes, and connections during training. However, as suugested by Angeline, Sauders and Pollack (1994) 'such *structural hill climbing* methods are susceptible to becoming trapped at structural local optima'. In addition, they 'only investigate restricted topological subsets rather than the complete class of network architectures'. (Angeline, Sauders and Pollack 1994).

Design of a near optimal CNN architecture can be formulated as a search problem in the architecture space where each point represents an architecture.

Given some performance [optimality] criterion for the architecture, e.g. minimum error, fastest learning, lowest complexity, etc., the performance level will form a surface in the space. The optimal architecture design is equivalent to finding the highest point on this surface. There are several characteristics of such a surface, as indicated by Miller, Todd and Hegde (1989), which make evolutionary algorithms better candidates for searching the surface than the constructive and pruning algorithms mentioned above.

This chapter describes a new evolutionary system, EPNet, for evolving feedforward CNNs (Yao and Liu 1997). It combines architectural evolution with weight learning. The evolutionary algorithm used to evolve CNNs is based on evolutionary programming [EP] (Fogel 1995a). It is argued in this chapter that EP is a better candidate than genetic algorithms [GAs] for evolving CNNs, since their emphasis on the behavioural link between parents and offspring can increase the efficiency of CNN's evolution.

EPNet is different from previous approaches to evolving CNNs for a number of reasons. *First,* EPNet emphasises the evolution of CNN behaviour (Yao 1997) and uses a number of techniques, such as partial training after each architectural mutation and node splitting, to effectively maintain the behavioural link between a parent and its offspring. While some of previous EP systems (Fogel, Fogel and Porto 1990; McDonnell and Waagen 1994; Angeline, Sauders and Pollack 1994; McDonnell and Waagen 1993; Fogel 1993; Fogel 1995a), acknowledged the importance of evolving behaviours, few techniques have been developed to maintain the behavioural link between parents and their offspring. The common practice in architectural mutations was to uniformly add or delete hidden nodes or connections at random. In particular, a hidden node was usually added to a hidden layer with full connections. Random initial weights were attached to these connections. Such an approach tends to destroy the behaviour already learned by the parent and create a poor behavioural link between the parent and its offspring.

Second, EPNet encourages parsimony of evolved CNNs by attempting different mutations sequentially. That is, node or connection deletion is always attempted before addition. If a deletion is 'successful', no other mutations will be made. Hence, a parsimonious CNN is always preferred. This approach is quite different from existing ones which add a network complexity [regularization] term in the fitness function to penalise large CNNs [i.e. the fitness function would look like $f = f_{error} + \alpha f_{complexity}$]. The difficulty in using such a function in practice lies in the selection of suitable coefficient α, which often involves tedious trial-and-error experiments. Evolving parsimonious CNNs by sequentially applying different mutations provides a novel and simple alternative which avoids the problem. The effectiveness of the approach has been demonstrated by the experimental results presented in this chapter.

Third, EPNet has been tested on a number of benchmark problems (Yao and Liu 1997; Liu and Yao 1996b), including the parity problem of various sizes, the Australian credit card assessment problem, four medical diagnosis problems [breast cancer, diabetes, heart disease, and thyroid], and the MacKey-Glass time series prediction problem. It was also tested on the two-spiral problem (Yao and Liu 1998b). Few evolutionary systems have been tested on a similar range of benchmark problems. The experimental results obtained by EPNet are better than those obtained by other systems in terms of generalization and the size of CNNs.

Although there have been many studies on how to evolve CNNs more effectively and efficiently (Yao 1991, 1993c, 1995, 1999), including the EPNet work, the issue of how to best use population information has been overlooked (Yao and Liu 1996a; Yao, Liu and Darwen 1996). Few attempts have been made to exploit population information. We argue in this chapter that learning is different from optimization. The difference between learning and optimization can be exploited in evolutionary computation where both learning and optimisation are population-based. It is shown that a population contains more information than any single individual in the population. Such information can be used to improve the generalization of learned systems. Four simple methods for combining different individuals in a population are described (Yao and Liu 1998a). Although the idea of combining different individuals in an ensemble has been studied in the CNN field and statistics (Hashem 1993; Perrone 1993), few attempts have been made in evolutionary learning to use population information to generate better systems.

The rest of this chapter is organized as follows. Section 3.2 begins by discussing different approaches to evolving CNN architectures and indicates potential problems with the existing approaches, then Section 3.3 describes EPNet in detail and gives motivations and ideas behind various design choices. In Section 3.4 experimental results on EPNet are presented and some aspects of these discussed. Section 3.5 examines the difference between evolutionary learning and optimisation, and the opportunity of exploiting this difference in population-based learning. Section 3.6 presents the four combination methods used in our studies. The purpose here is not to find the best combination method, but to demonstrate the effectiveness of our approach even with some simple methods. Finally, Section 3.7 concludes with a summary of the chapter and a few remarks.

3.2 Evolving Computational Neural Network Architectures

There are two major approaches to evolving CNN architectures (Yao 1991, 1999). One is the evolution of 'pure' architectures [i.e., architectures without weights]. Connection weights will be trained after a near optimal architecture has been found. The other is the simultaneous evolution of both architectures and weights.

The Evolution of Pure Architectures

One major issue in evolving pure architectures is to decide how much information about an architecture should be encoded into a chromosome [genotype]. At one extreme, all the detail – every connection and node – can be specified by the genotype, for example by some binary bits. This kind of representation scheme is called a *direct encoding scheme* or *strong specification scheme*. At the other extreme, only the most important parameters of an architecture, such as the number of hidden layers and hidden nodes in each layer are encoded. Other details are either pre-defined or left to the training process to decide. This kind of representation scheme is called a *indirect encoding scheme* or *weak specification scheme*. Fig. 3.1 (Yao 1993b, 1995) shows the evolution of pure architectures under either a direct or an indirect encoding scheme.

(i) Decode each individual [i.e., chromosome] in the current generation into an architecture. If the indirect encoding scheme is used, further detail of the architecture is specified by some developmental rules or a training process.
(ii) Train each neural network with the decoded architecture by a pre-defined learning rule/algorithm [some parameters of the learning rule could be learned during training] starting from different sets of random initial weights and, if any, learning parameters.
(iii) Define the fitness of each individual [encoded architecture] according to the above training result and other performance criteria such as the complexity of the architecture.
(iv) Reproduce a number of children for each individual in the current generation based on its fitness.
(v) Apply genetic operators to the children generated above and obtain the next generation.

Fig. 3.1 A typical cycle in the evolution of architectures

It is worth pointing out that genotypes in Fig. 3.1 do not contain any weight information. In order to evaluate them, they have to be trained from a random set of initial weights using a training algorithm like BP. Unfortunately, such fitness evaluation of the genotypes is very noisy because a phenotype's fitness is used to represent the genotype's fitness. There are two major sources of noise:

(i) The *first source* is the random initialization of the weights. Different random initial weights may produce different training results. Hence, the same genotype may have quite different fitness due to different random initial weights used by the phenotypes [for more details, see Chapter 8 in this volume].
(ii) The *second source* is the training algorithm. Different training algorithms may produce different training results even from the same set of initial weights. This is especially true for multimodal error functions. For example, a BP may reduce a CNN's error to 0.05 through training, but an EP could reduce the error to 0.001 due to its global search capability.

Such noise can mislead the evolution because the fact that the fitness of a phenotype generated from genotype G_1 is higher than that generated from genotype G_2 does not mean that G_1 has higher fitness than G_2. In order to reduce such noise, an architecture usually has to be trained many times from different random initial weights. The average results will then be used to estimate the genotype's fitness. This method increases the computation time for fitness evaluation dramatically. It is one of the major reasons why only small CNNs were evolved in previous studies (Whitley and Starkweather 1990b; Yao and Shi 1995).

In essence, the noise identified in this chapter is caused by the one to many mapping from genotypes to phenotypes. Angeline, Sauders and Pollack (1994) and Fogel (1995a, b) have provided a more general discussion on the mapping between genotypes and phenotypes. It is clear that the evolution of pure architectures has difficulty in evaluating fitness accurately. As a result, the evolution would be very inefficient.

The Simultaneous Evolution of Both Architectures and Weights

One way to alleviate the noisy fitness evaluation problem is to have a one to one mapping between genotypes and phenotypes. That is, both architecture and weight information are encoded in individuals and are evolved simultaneously. Although the idea of evolving both architectures and weights is not new (Angeline, Sauders and Pollack 1994; Maniezzo 1994; McDonnell and Waagen 1994; Fogel 1995a), few have explained why it is important in terms of accurate fitness evaluation. The simultaneous evolution of both architectures and weights is summarized in Fig. 3.2. The evolution of CNN architectures in general suffers from the permutation problem (Belew, McInerney and Schraudolph 1991) or so-called competing conventions problem. It is caused by the many to one mapping from genotypes to phenotypes, since two CNNs which order their hidden nodes differently may have different genotypes but are behaviourally [i.e., phenotypically] equivalent. This problem not only makes the evolution inefficient, but also makes crossover operators more difficult to produce highly fit offspring. It is unclear what building blocks actually are in this situation.

Some work has been undertaken relating to evolving CNN architectures. For example, Smalz and Conrad (1994) proposed a novel approach to assigning credits and fitness to neurons [i.e., nodes or processing elements] in an CNN, rather than the CNN itself. This is quite different from all other methods which only evaluate a complete CNN without going inside it. The idea is to identify those neurons which 'are most compatible with all of the network contexts associated with the best performance on any of the inputs' (Smalz and Conrad 1994). Starting from a population of redundant, identically structured networks that vary only with respect to individual neuron parameters, their evolutionary method first evaluates neurons and then copies with mutation the parameters of those neurons that have high fitness values to other processing elements in the same class. In other words, it tries to put all fit neurons together to hopefully generate a fit network. However, Smalz and Conrad's evolutionary method does not change the network architecture, which is fixed (Smalz and Conrad 1994). The appropriateness of assigning credit/fitness to individual neurons also needs further investigation. It is well-known that CNNs use distributed representation. It is difficult to identify a single neuron for the good or poor performance of a network. Putting a group of 'good' neurons from different CNNs together may not produce a better CNN unless a local representation is used. It appears that Smalz and Conrad's (1994) method is best suited to CNNs such as radial basis function [RBF] networks.

(vi) *Evaluate each individual based on their errors and/or other performance criteria such as its complexity.*
(vii) *Select individuals for reproduction and genetic operation.*
(viii) *Apply genetic operators, such as crossover and mutation, to the CNN architectures and weights, and obtain the next generation.*

Note: The term 'genetic' used above is rather loose and should not be interpreted in the strict biological sense. Genetic operators are just search operators.

Fig. 3.2 A typical cycle of the evolution of both architectures and weights

Odri, Petrovacki and Krstonosic (1993) proposed a non-population based learning algorithm which could change CNN architectures. It uses the idea of evolutional development. The algorithm is based on BP. During training, a new neuron may be added to the existing CNN through 'cell division' if an existing neuron generates a non-zero error (Odri, Petrovacki and Krstonosic 1993). A connection may be deleted if it does not change very much in previous training steps. A neuron is deleted only when all of its incoming or all of its outgoing connections have been deleted. There is no obvious way to add a single connection. The algorithm was only tested on the XOR problem to illustrate its basic features (Odri, Petrovacki and Krstonosic 1993). One major disadvantage of this algorithm is its tendency to generate larger-than-necessary CNN and overfit training data. It can only deal with strictly layered CNNs.

3.3 EPNet

In order to reduce the detrimental effect of the permutation problem, EPNet adopts an EP algorithm which does not use crossover. EP's emphasis on the behavioural link between parents and their offspring also matches well with the emphasis on evolving CNN behaviours, not just circuitry. In its current implementation, EPNet is used to evolve feedforward CNNs with sigmoid transfer functions. However, this is not an inherent constraint. In fact, EPNet has minimal constraint on the type of CNNs which may be evolved. The feedforward CNNs do not have to be strictly layered or fully connected between adjacent layers. They may also contain hidden nodes with different transfer functions (Liu and Yao 1996a).

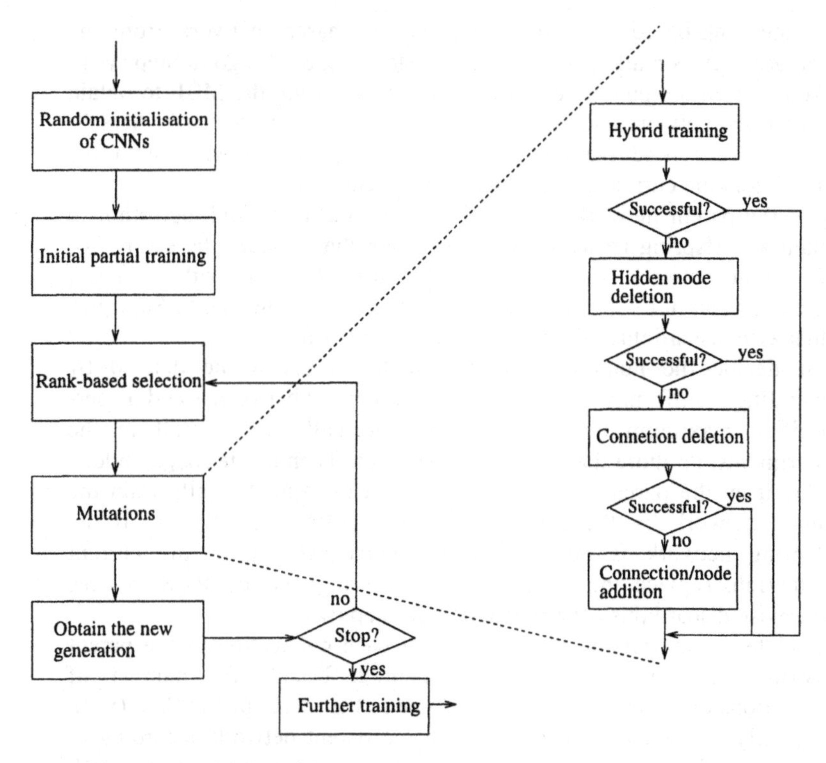

Fig. 3.3 Major steps of EPNet

The major steps of EPNet are described by Fig. 3.3, which are explained further as follows (Yao and Liu 1997, 1998b):

(i) Generate an initial population of M networks at random. The number of hidden nodes and the initial connection density for each network are uniformly generated at random within certain ranges. The random initial weights are uniformly distributed inside a small range.

(ii) Partially train each network in the population on the training set for a certain number of epochs using a modified BP [MBP] with adaptive learning rates. An epoch represents one sweep of the whole training data. The number of epochs, K_0, is specified by the user. The error value E of each network on the validation set is checked after partial training. If E has not been significantly reduced, then the assumption is that the network is trapped in a local minimum and the network is marked 'failure'. Otherwise the network is marked 'success'.

(iii) Rank the networks in the population according to their error values, from the best to the worst.

(iv) If the best network found is acceptable or the maximum number of generations has been reached, stop the evolutionary process and go to Step (xi). Otherwise continue.

(v) Use the rank-based selection to choose one parent network from the population. If its mark is 'success', go to Step (vi), or else go to Step (vii).

(vi) Partially train the parent network for K_1 epochs using the MBP to obtain an offspring network and mark it in the same way as in Step (ii), where K_1 is a user specified parameter. Replace the parent network with the offspring in the current population and go to Step (iii).

(vii) Train the parent network with a simulated annealing [SA] algorithm to obtain an offspring network. If the SA algorithm reduces the error E of the parent network significantly, mark the offspring with 'success', replace its parent by it in the current population, and then go to Step (iii). Otherwise discard this offspring and go to Step (viii).

(viii) First decide the number of hidden nodes N_{hidden} to be deleted by generating a uniformly distributed random number between 1 and a user-specified maximum number. N_{hidden} is normally very small in the experiments, no more than three in most cases. Then delete N_{hidden} hidden nodes from the parent network uniformly at random. Partially train the pruned network by the MBP to obtain an offspring network. If the offspring network is better than the worst network in the current population, replace the worst by the offspring and go to Step (iii). Otherwise discard this offspring and go to Step (ix).

(ix) Calculate the approximate importance of each connection in the parent network using the nonconvergent method. Decide the number of connections to be deleted in the same way as that described in Step (viii). Randomly delete the connections from the parent network according to the calculated importance. Partially train the pruned network by the MBP to obtain an offspring network. If the offspring network is better than the worst network in the current population, replace the worst by the offspring and go to Step (iii). Otherwise discard this offspring and go to Step (x).

(x) Decide the number of connections and nodes to be added in the same way as that described in Step (viii). Calculate the approximate importance of each virtual connection with zero weight. Randomly add the connections to the parent network to obtain *Offspring 1* according to their importance. Addition of each node is implemented by splitting a randomly selected hidden node in the parent network. The new grown network after adding all nodes is *Offspring 2*. Partially train *Offspring 1* and *Offspring 2* by the MBP to obtain a survival offspring. Replace the worst network in the current population by the offspring and go to Step (iii).

(xi) After the evolutionary process, train the best network further on the combined training and validation set until it 'converges'.

The above evolutionary process appears to be rather complex, but in essence it is an EP algorithm with five mutations: hybrid training, node deletion, connection deletion, connection addition and node addition. Details about each component of EPNet are given in the following subsections.

Encoding Scheme for Feedforward Computational Neural Networks

The feedforward CNNs considered by EPNet are generalized multilayer perceptrons (Werbos 1994, pp. 272-273). The architecture of such networks is shown in Fig. 3.4, where **X** and **Y** are inputs and outputs respectively.

$$x_i = X_i \qquad \text{with} \quad 1 \le i \le m \tag{3.1}$$

$$net_i = \sum_{j=1}^{i-1} w_{ij}\, x_j \qquad \text{with} \quad m < i \le m + N + n \tag{3.2}$$

$$x_j = f(net_j) \qquad \text{with} \quad m < j \le m + N + n \tag{3.3}$$

$$Y_i = x_{i+m+N} \qquad \text{with} \quad 1 \le i \le n \tag{3.4}$$

where f is the following sigmoid function:

$$f(z) = (1 + \exp(-z))^{-1} \tag{3.5}$$

m and n are the number of inputs and outputs respectively, N is the number of hidden nodes.

In Fig. 3.4, there are m+N+n circles, representing all of the nodes in the network, including the input nodes. The first m circles are really just copies of the inputs X_1, ..., X_m. Every other node in the network, such as node number i, which calculates net_i and x_i, takes inputs from every node that precedes it in the network. Even the last output node [the (m+N n)th], which generates Y_n, takes input from other output nodes, such as the one which outputs Y_{n-1}.
The direct encoding scheme is used in EPNet to represent CNN architectures and connection weights [including biases]. This is necessary because EPNet evolves CNN architectures and weights simultaneously and needs information about every connection in an CNN. Two equal size matrices and one vector are used to specify an CNN in EPNet. The dimension of the vector is determined by a user-specified upper limit N, which is the maximum number of hidden nodes allowable in the CNN. The size of the two matrices is (m+N+n) (m+N+n), where m and n are the number of input and output nodes respectively. One matrix is the connectivity matrix of the CNN, whose entries can only be 0 or 1. The other is the corresponding weight matrix whose entries are real numbers. The decision to use two matrices rather than one is purely implementation-driven. The entries in the hidden node vector can be either 1, i.e. the node exists, or 0, i.e. the node does not exist. Since this chapter is only concerned with feedforward CNNs, only the upper triangle will be considered in the two matrices.

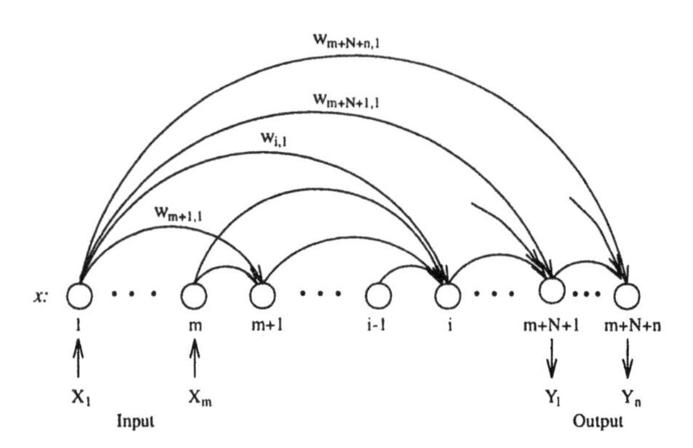

Fig. 3.4 A fully connected feedforward computational neural network (Werbos 1994, pp. 273)

Fitness Evaluation and Selection Mechanism

The fitness of each individual in EPNet is solely determined by the inverse of an error value defined by Equation (3.6) (Prechelt 1994) over a *validation* set containing T patterns:

$$E = 100\frac{o_{\max} - o_{\min}}{Tn}\sum_{t=1}^{T}\sum_{i=1}^{n}(Y_i(t) - Z_i(t))^2 \tag{3.6}$$

where o_{max} and o_{min} are the maximum and minimum values of output coefifcients in the problem representation, n is the number of output nodes, $Y_i(t)$ and $Z_i(t)$ are actual and desired outputs of node i for pattern t.

Equation (3.6) was suggested by Prechelt (1994) to make the error measure less dependent on the size of the validation set and the number of output nodes. A mean squared error percentage was therefore adopted. o_{max} and o_{min} were the maximum and minimum values of outputs (Prechelt 1994).

The fitness evaluation in EPNet is different from previous work in ECNNs [evolutionary computational neural networks], since it is determined through a validation set which does not overlap with the training set. Such use of a validation set in an evolutionary learning system improves the generalization ability of evolved CNNs and introduces few overheads in computation time.

The selection mechanism used in EPNet is rank based. Let M sorted individuals be numbered as 0, 1, ..., $M-1$ with the 0th being the fittest. Then the $(M-j)$th individual is selected with probability (Yao 1993a)

$$p(M-j)=\frac{j}{\sum_{k=1}^{M} k} \tag{3.7}$$

The selected individual is then modified by the five mutations. In EPNet, error E is used to sort individuals directly rather than to compute $f = 1/E$ and use f to sort them.

Replacement Strategy and Generation Gap

The replacement strategy used in EPNet reflects the emphasis on evolving CNN behaviours and maintaining behavioural links between parents and their offspring. It also reflects the fact that EPNet actually emulates a kind of Lamarckian rather than Darwinian evolution. There is on-going debate as to whether Lamarckian evolution or the Baldwin effect is more efficient in simulated evolution (Ackley and Littman 1994; Whitley, Gordon and Mathias 1994). Ackley and Littman (1994) have presented a case for Lamarckian evolution. The experimental results of EPNet seem to support their view.

In EPNet, if an offspring is obtained through further BP partial training, it always replaces its parent. If an offspring is obtained through SA training, it replaces its parent only when this reduces its error significantly. If an offspring is obtained through deleting nodes/connections, it replaces the worst individual in the population only when it is better than the worst. If an offspring is obtained through adding nodes/connections, it always replaces the worst individual in the population, since an CNN with more nodes/connections is more powerful, although its current performance may not be very good due to incomplete training.

The generation gap in EPNet is minimal. That is, a new generation starts immediately after the above replacement. This is very similar to the steadystate GA (Whitley and Starkweather 1990a; Syswerda 1991) and continuous EP (Fogel and Fogel 1995), although the replacement strategy used in EPNet is different.

Hybrid Training

The only mutation for modifying CNN's weights in EPNet is implemented by a hybrid training algorithm consisting of an MBP and a simplified SA algorithm.

The classical BP algorithm (Rumelhart, Hinton and Williams 1986) is notorious for its slow convergence and convergence to local minima. Hence it is modified in order to alleviate these two problems. A simple heuristic is used to adjust the learning rate for each CNN in the population. Different CNNs may have different learning rates. During BP training, the error E is checked after every k epochs, where k is a parameter determined by the user. If E decreases, the learning rate is increased by a predefined amount. Otherwise, the learning rate is reduced. In the later case the new weights and error are discarded.

In order to deal with the local optimum problem suffered by the classical BP algorithm, an extra training stage is introduced when BP training cannot improve a CNN anymore. The extra training is performed by an SA algorithm. When the SA algorithm also fails to improve the CNN, the four mutations will be used to change the CNN architecture. It is important in EPNet to train a CNN first without modifying its architecture. This reflects the emphasis on a close behavioural link between the parent and its offspring.

The hybrid training algorithm used in EPNet is not a critical choice in the whole system. Its main purpose is to discourage architectural mutations if training, which often introduces smaller behavioural changes in comparison with architectural mutations, can produce a satisfactory CNN. Other training algorithms which are faster and can avoid poor local minima can also be used (Yao and Liu 1996b, Yao, Liu and Lin 1999). The investigation of the best training algorithm is outside the scope of this chapter.

Architecture Mutations

In EPNet, only when the hybrid training fails to reduce the error of an CNN will architectural mutations take place. For architectural mutations, node or connection deletions are always attempted before connection or node additions in order to encourage the evolution of small CNNs. Connection or node additions will be tried only after node or connection deletions fail to produce a good offspring. Using the order of mutations to encourage parsimony of evolved CNNs represents a dramatically different approach from using a complexity [regularization] term in the fitness function [see Chapter 8 in this volume]. It avoids the time-consuming trial-and-error process of selecting a suitable coefficient for the regularization term.

Hidden Node Deletion. Certain hidden nodes are first deleted uniformly at random from a parent CNN. The maximum number of hidden nodes that can be deleted is set by a user-specified parameter. Then the mutated CNN is partially trained by the MBP. This extra training process can reduce the sudden behavioural change caused by the node deletion. If this trained CNN is better than the worst CNN in the population, the worst CNN will be replaced by the trained one and no further mutation will take place. Otherwise connection deletion will be attempted.

Connection Deletion. Certain connections are selected probabilistically for deletion according to their importance. The maximum number of connections that can be deleted is set by a user-specified parameter. The importance is defined by a significance test for the weight's deviation from zero in the weight update process (Finnoff, Hergent and Zimmermann 1993).

Denote the weight update $\Delta_{ij}(w) = -\eta[\partial L_t / \partial w_{ij}]$ by the local gradient of the linear error function L ($L = \Sigma_{t=1}^{T} \Sigma_{i=1}^{n} |Y_i(t) - Z_i(t)_j|$) with respect to example t and weight w_{ij}, the significance of the deviation of w_{ij} from zero is defined by the test variable (Finnoff, Hergent und Zimmermann 1993)

$$test\,(w_{ij}) = \frac{\sum_{t=1}^{T} \xi_{ij}^{t}}{\sqrt{\sum_{t=1}^{T} (\xi_{ij}^{t} - \overline{\xi}_{ij})^2}} \tag{3.8}$$

where $\xi_{ij}^{t} = w_{ij} + \Delta w_{ij}^{t}(w)$, $\overline{\xi}_{ij}$ denotes the average over the set ξ_{ij}^{t}, $t = 1, ..., T$. A large value of test variable test (w_{ij}) indicates higher importance of the connection with weight w_{ij}.

The advantage of the above nonconvergent method (Finnoff, Hergent und Zimmermann 1993) over others is that it does not require the training process to converge in order to test connections. It does not require any extra parameters either. For example, Odri, Petrovacki and Krstonosic's (1993) method needs to 'guess' values for four additional parameters. The idea behind the test variable (3.8) is to test the significance of the deviation of w_{ij} from zero (Finnoff, Hergent und Zimmermann 1993). Equation (3.8) can also be used for connections whose weights are zero, and thus can be used to determine which connections should be added in the addition phase.

Similar to the case of node deletion, the CNN will be partially trained by the MBP after certain connections have been deleted from it. If the trained CNN is better than the worst CNN in the population, the worst CNN will be replaced by the trained one and no further mutation will take place. Otherwise node/connection addition will be attempted.

Connection and Node Addition. As mentioned before, certain connections are added to a parent network probabilistically according to Equation (3.8). They are selected from those connections with zero weights. The added connections are initialized with small random weights. The new CNN will be partially trained by the MBP and denoted as *Offspring 1.*

Node addition is implemented through splitting an existing hidden node, a process called 'cell division' by Odri, Petrovacki and Krstonosic (1993). In addition to reasons given by Odri, Petrovacki and Krstonosic (1993), growing an CNN by splitting existing ones can preserve the behavioural link between the parent and its offspring better than by adding random nodes. The nodes for splitting are selected uniformly at random among all hidden nodes. Two nodes obtained by splitting an existing node i have the same connections as the existing node. The weights of these new nodes have the following values (Odri, Petrovacki and Krstonosic 1993):

$$w_{ij}^{1} = w_{ij}^{2} = w_{ij} \qquad \text{for } i \geq j \tag{3.9}$$

$$w_{ki}^{1} = (1 + \alpha)\, w_{ki} \qquad \text{for } i < k \tag{3.10}$$

$$w_{ki}^{2} = -\,\alpha\, w_{ki} \qquad \text{for } i < k \tag{3.11}$$

where $\boldsymbol{w}$ is the weight vector of the existing node i, $\boldsymbol{w}^1$ and $\boldsymbol{w}^2$ are the weight vectors of the new nodes, and α is a mutation parameter which may take either a fixed or random value. The split weights imply that the offspring maintains a strong behavioural link with the parent. For training examples which were learned correctly by the parent, the offspring needs little adjustment of its inherited weights during partial training.

The new CNN produced by node splitting is denoted as *Offspring 2.* After it is generated, it will also be partially trained by the MBP. Then it has to compete with *Offspring 1* for survival. The survived one will replace the worst CNN in the population.

Further Training After Evolution

One of the most important goals for CNNs is to have a good generalization ability. In EPNet, a training set is used for the MBP and a validation set for fitness evaluation in the evolutionary process. After the simulated evolution, the best evolved CNN is further trained using the MBP on the combined training and validation set. Then this further trained CNN is tested on an unseen testing set to evaluate its performance.

Alternatively, all the CNNs in the final population can be trained using the MBP and the one which has the best performance on a second validation set is selected as EPNet's final output. This method is more time-consuming, but it considers all the information in the final population rather than just the best individual. The importance of making use of the information in a population has recently been demonstrated by evolving both CNNs (Yao and Liu 1996a; Yao, Liu and Darwen 1996) and rule-based systems (Darwen and Yao 1996a; Yao, Liu and Darwen 1996). The use of a second validation set also helps to prevent CNNs from overfitting the combined training and the first validation set. Experiments using either one or two validation sets will be described in the following section.

3.4 Experimental Studies

EPNet was first tested on the *N parity problem* where $N = 4 - 8$ (Liu and Yao 1996b). The results obtained by EPNet (Yao and Liu 1997) are quite competitive in comparison with those obtained by other algorithms. Table 3.1 compares EPNet's best results with those of cascade-correlation algorithm [CCA] (Fahlman and Lebiere 1990), the perceptron cascade algorithm [PCA] (Burgess 1994), the tower algorithm [TA] (Nadal 1989), and the FNNCA (Setiono and Hui 1995). All these algorithms, except for the FNNCA, can produce networks with short cut connections. Two observations can be made from this table. *First,* EPNet can evolve very compact networks. In fact, it generated the smallest CNN among the five algorithms compared here. *Second,* the size of the network evolved by EPNet seems to grow more slowly than that produced by other algorithms when the size of the problem [i.e. *N*] increases. That is, EPNet seems to perform even better for large problems in terms of the number of hidden nodes. Since CCA, PCA and TA are all fully connected, the number of connections in EPNet-evolved CNNs is smaller as well.

Table 3.1 Comparison between EPNet and other algorithms in terms of the minimal number of hidden nodes in the best network generated

Algorithm	EPNet	CCA	PCA	TA	FNNCA
Hidden Nodes	2,2,3,3,3	2,2,3,4,5	2,2,3,3,4	2,2,3,3,4	3,4,5,5,-

Notes: The 5-tuples in the table represent the number of hidden nodes for the 4, 5, 6, 7, and 8 parity problem respectively. '-' means no result is available.

Fig. 3.5 shows the best network evolved by EPNet for the 8 parity problem. Table 3.2 gives its weights. It is rather surprising that a 3 hidden node network can be found by EPNet for the 8 parity problem. This demonstrates an important point made by many evolutionary algorithm researchers – an evolutionary algorithm can often discover novel solutions which are very difficult for human beings to find.

Although Stork and Allen report a two hidden node CNN which can solve the *N* parity problem (Stork and Allen 1992), their network was hand-crafted and used a very special node transfer function rather than the usual sigmoid one.

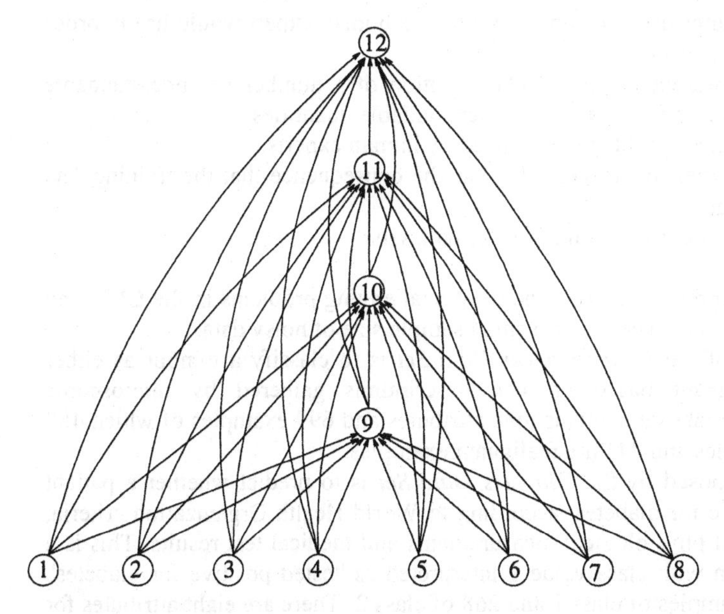

Fig. 3.5 The best network evolved by EPNet for the 8 parity problem

Table 3.2 Connection weights and biases [represented by *T*] for the network in Fig. 3.5

	T	1	2	3	4	5
9	–12.4	25.2	27.7	–29.4	–28.9	–29.7
10	–40.4	19.6	18.9	–18.1	–19.1	–18.5
11	–48.1	16.0	16.1	–15.9	–16.3	–15.8
12	45.7	–10.0	–11.0	10.0	9.9	9.4
	6	**7**	**8**	**9**	**10**	**11**
9	–25.4	–28.5	27.8	0.0	0.0	0.0
10	–17.3	–18.8	20.4	–67.6	0.0	0.0
11	–15.9	–15.8	16.7	–55.0	–26.7	0.0
12	10.0	9.6	–11.4	6.8	2.3	76.3

Note: The numbers refer to the nodes of the CNN as visualized in Fig. 3.5.

The Medical Diagnosis Problems

Since the training set was the same as the testing set in the experiments with the *N* parity problem, EPNet was only tested for its ability to evolve CNNs that *learn* well, but not necessarily *generalize* well. In order to evaluate EPNet's ability in evolving CNNs that generalize well, EPNet was applied to four real-world problems in the medical domain, i.e. diagnosis problems concerning breast cancer, diabetes, heart disease, and the thyroid. All date sets were obtained from the UCI machine learning benchmark repository. These medical diagnosis problems have the following common characteristics (Prechelt 1994):

- The input attributes used are similar to those a human expert would use in order to solve the same problem.
- The outputs represent either the classification of a number of understandable classes or the prediction of a set of understandable quantities.
- In practice, all these problems are solved by human experts.
- Examples are expensive to get. This has the consequence that the training sets are not very large.
- There are missing attribute values in the data sets.

These data sets represent some of the most challenging problems in the CNN and machine learning field. They have a small sample size of noisy data.

The purpose of the *Breast Cancer Data Set* is to classify a tumour as either benign or malignant based on cell descriptions gathered by microscopic examination. The data set contains nine attributes and 699 examples of which 458 are benign examples and 241 are malignant examples.

The problem posed by the *Diabetes Data Set* is to predict whether a patient would test positive for diabetes, according to World Health Organization criteria, given a number of physiological measurements and medical test results. This is a two class problem with class value 1 interpreted as 'tested positive for diabetes'. There are 500 examples of class 1 and 268 of class 2. There are eight attributes for each example. The data set is rather difficult to classify. The so-called 'class' value is really a binarized form of another attribute, which is itself highly indicative of certain types of diabetes, but does not have a one to one correspondence with the medical condition of being diabetic.

The purpose of the *Heart Disease Data Set* is to predict the presence or absence of heart disease given the results of various medical tests carried out on a patient. This database contains 13 attributes, which have been extracted from a larger set of 75. The database originally contained 303 examples, but six of these contained missing class values and so were discarded, leaving 297. 27 of these were retained in case of dispute, leaving a final total of 270. There are two classes: presence and absence [of heart disease]. This is a reduction of the number of classes in the original data set in which there were four different degrees of heart disease.

The *Thyroid Data Set* comes from the 'ann' version of the thyroid disease data set from the UCI machine learning repository. Two files were provided: 'anntrain.data' contains 3,772 learning examples and 'ann-test.data' contains 3,428 testing examples. There are 21 attributes for each example. The purpose of the data set is to determine whether a patient referred to the clinic is hypothyroid. Therefore three classes are built: normal [not hypothyroid], hyperfunction and subnormal functioning. Because 92 percent of the patients are not hyperthyroid, a good classifier must be significantly better than 92 percent.

Experimental Set-Up. All the data sets used by EPNet were partitioned into three sets: a training set, a validation set, and a testing set. The training set was used to train CNNs by MBP, the validation set was used to evaluate the fitness of the CNNs. The best CNN evolved by EPNet was further trained on the combined training and validation set before it was applied to the testing set.

As indicated by Prechelt (1994, 1995), it is insufficient to indicate only the number of examples for each set in the above partition, because the experimental results may vary significantly for different partitions, even when the numbers in each set are the same [see Chapter 8 in this volume for more details]. An imprecise specification of the partition of a known data set into the three sets is one of the most frequent obstacles to reproducing and comparing published neural network learning results. In the following experiments, each data set was partitioned as follows:

- For the breast cancer data set, the first 349 examples were used for the training set, the following 175 examples for the validation set, and the final 175 examples for the testing set.
- For the diabetes data set, the first 384 examples were used for the training set, the following 192 examples for the validation set, the final 192 examples for the testing set.
- For the heart disease data set, the first 134 examples were used for the training set, the following 68 examples for the validation set, and the final 68 examples for the testing set.
- For the thyroid data set, the first 2,514 examples in 'ann-train.data' were used for the training set, the rest of 'ann-train.data' were used for the validation set, and the whole 'ann-test.data' for the testing set.

The input attributes of the diabetes data set and heart disease data set were rescaled to between 0.0 and 1.0 by a linear function. The output attributes of all the problems were encoded using a 1-of-*m* output representation for *m* classes. The winner-takes-all method was used in EPNet, i.e. the output with the highest activation designates the class.

There are some control parameters in EPNet which need to be specified by the user. It is however unnecessary to tune all these parameters for each problem because EPNet is not very sensitive to them. Most parameters used in the experiments were set to be the same: the population size [20], the initial connection density [1.0], the initial learning rate [0.25], the range of learning rate [0.1 to 0.75], the number of epochs for the learning rate adaptation [5], the number of mutated hidden nodes [1], the number of mutated connections [1 to 3], the number of temperatures in SA [5], and the number of iterations at each temperature [100]. The different parameters were the number of hidden nodes of each individual in the initial population and the number of epochs for MBP's partial training. The number of hidden nodes for each individual in the initial population was chosen from a uniform distribution within certain ranges: 1 to 3 hidden nodes for the breast cancer problem; 2 to 8 for the diabetes problem; 3 to 5 for the heart disease problem; and 6 to 15 for the thyroid problem.

The number of epochs [K_0] for training each individual in the initial population is determined by two user-specified parameters: the 'stage' size and the number of stages. A stage includes a certain number of epochs for MBP's training. The two parameters mean that a CNN is first trained for one stage. If the error of the network reduces, then another stage is executed, or else the training finishes. This step can repeat up to *the-number-of-stages* times. This simple method balances fairly well between the training time and the accuracy. For the breast cancer problem and the diabetes problem, the two parameters were 400 and 2. For the heart disease problem, they were 500 and 2. For the thyroid problem, they were 350 and 3.

The number of epochs for each partial training during evolution [i.e. K_1] was determined in the same way as the above. The two parameters were 50 and 3 for the thyroid problem, 100 and 2 for the other problems. The number of epochs for training the best individual on the combined training and testing data set was set to be the same [1,000] for all four problems. A run of EPNet was terminated if the average error of the population had not decreased by more than a threshold value ε after consecutive G_0 generations or a maximum number of generations was reached. The same maximum number of generations [500] and the same G_0 [10] were used for all four problems. The threshold value ε was set to 0.1 for the thyroid problem, and 0.01 for the other three. These parameters were chosen after some limited preliminary experiments. They were not meant to be optimal.

Experimental Results. Table 3.3 and Table 3.4 show EPNet's results over 30 runs. The *error* in the tables refers to the error defined by Equation (3.6). The *error rate* refers to the percentage of wrong classifications produced by the evolved CNNs. It is clear from the two tables that the evolved CNNs have very small sizes, i.e., a small number of hidden nodes and connections, as well as low error rates. For example, an evolved CNN with just one hidden node can achieve an error rate of 19.794 percent on the testing set for the diabetes problem. Another evolved CNN with just three hidden nodes can achieve an error rate of 1.925 percent on the testing set for the thyroid problem.

Table 3.3 Architectures of evolved computational neural networks

		Number of Connections	Number of Hidden Nodes	Number of Generations
Breast Cancer Data Set	Mean	41.0	2.0	137.3
	Standard Deviation	14.7	1.1	37.7
	Minimum	15.0	0.0	100.0
	Maximum	84.0	5.0	240.0
Diabetes Data Set	Mean	52.3	3.4	132.2
	SD	16.1	1.3	48.7
	Minimum	27.0	1.0	100.0
	Maximum	87.0	6.0	280.0
Heart Data Set	Mean	92.6	4.1	193.3
	SD	40.8	2.1	60.3
	Minimum	34.0	1.0	120.0
	Maximum	213.0	10.0	320.0
Thyroid Data Set	Mean	219.6	5.9	45.0
	SD	74.4	2.4	12.5
	Minimum	128.0	3.0	10.0
	Maximum	417.0	12.0	70.0

Table 3.4 Accuracies of evolved computational neural networks, where 'E/R' indicates error rate and 'M/n' indicates mean

		Training Set		Validation Set		Test Set	
		Error	**E/R**	**Error**	**E/R**	**Error**	**E/R**
Breast Cancer	Mean	3.246	0.03773	0.644	0.00590	1.421	0.01376
	SD	0.589	0.00694	0.213	0.00236	0.729	0.00938
	Min	1.544	0.01719	0.056	0.00000	0.192	0.00000
	Max	3.890	0.04585	1.058	0.01143	3.608	0.04000
Diabetes	Mean	16.674	0.24054	13.308	0.18854	15.330	0.22379
	SD	0.294	0.00009	0.437	0.00008	0.300	0.00014
	Min	16.092	0.21875	12.574	0.16667	14.704	0.19271
	Max	17.160	0.26042	14.151	0.20313	15.837	0.25000
Heart	Mean	10.708	0.13632	13.348	0.17304	12.270	0.16765
	SD	0.748	0.01517	0.595	0.01995	0.724	0.02029
	Min	8.848	0.09702	12.388	0.13235	10.795	0.13235
	Max	12.344	0.16418	14.540	0.20588	14.139	0.19118
Thyroid	Mean	0.470	0.00823	0.689	0.01174	1.157	0.02115
	SD	0.091	0.00146	0.127	0.00235	0.098	0.00220
	Min	0.336	0.00517	0.469	0.00636	0.887	0.01634
	Max	0.706	0.01154	1.066	0.01749	1.328	0.02625

Comparisons with Other Work. Direct comparison with other evolutionary approaches to designing CNNs is very difficult due to the lack of such results. Instead, the best and latest results available in the literature, regardless of whether the algorithm used was an evolutionary, a BP or a statistical one, were used in the comparison. It is possible that some papers which should have been compared were overlooked. However, the aim here is not to compare EPNet exhaustively with all other algorithms, but to present the underlying ideas.

Setiono and Hui (1995) have recently published a new CNN constructive algorithm called FNNCA. Prechelt (1994) also reported results on manually constructed CNNs. He tested a number of different CNN architectures for the breast cancer problem. The *best* results produced by FNNCA (Setiono and Hui 1995) and by hand-designed CNNs [denoted as HDCNNs] (Prechelt 1994) are compared to the *average* results produced by EPNet in Table 3.5.

Although EPNet can evolve very compact CNNs which generalize well, they come with the cost of additional computation time needed to perform search. The total time used by EPNet could be estimated by adding the initial training time [400.20 = 8,000 epochs], the evolving time [approximately 200 epochs per generation for maximally 500 generations], and the final training time [1,000 epochs] together. That is, it could require roughly 109,000 epochs for a single run. The actual time was less, since few runs reached the maximal number of generations. Similar estimations can be applied to other problems tested in this chapter. For many applications, the training time is less important than the generalization. Section 3.1 has explained why the evolutionary approach is necessary and better than constructive algorithms for such applications.

Table 3.5 Comparison between FNNCA, a hand-designed CNN and EPNet on the breast cancer problem

	Best Results FNNCA[a] 50 Runs	**Best Results HDCNNS[b] Trial-and-Error**	**Average Results EPNet 30 Runs**
Hidden Nodes	2	6	2.0
Testing E/R[c]	0.0145[d]	0.11149	0.01376

Notes: **a** FNNCA is a new CNN constructive algorithm (Setione and Hui 1995).
b HDCNNS is a hand-designed CNN (Prechelt 1994).
c E/R indicates the error rate.
d This minimum error rate was achieved when the number of hidden nodes was 3. The testing error rate was 0.0152 when the number of hidden nodes was 2.
CNNs designed manually and by FNNCA have more connections than those evolved by EPNet, even when the number of hidden nodes is the same, since EPNet can generate sparsely connected CNNs. Only the average results from EPNet are shown here. EPNet's best results are clearly superior, as indicated by Table 3.4.

The *diabetes problem* is one of the most challenging problems in CNN and machine learning due to its relatively small data set and high noise level. In the medical domain, data are often very costly to obtain. It would be unreasonable if an algorithm relied on more training data to improve its generalization. Table 3.6 compares EPNet's result with those produced by a number of other algorithms (Michie, Spiegelhalter and Taylor 1994). It is worth pointing out that the other results were obtained by 12-fold cross validation (Michie, Spiegelhalter and Taylor 1994). They represented the best 11 out of 23 algorithms tested.

In terms of best results produced, Prechelt (1994) tried different CNNs manually for the problem and found an 8 hidden node CNN which achieved the testing error rate of 0.2135 [21.35 percent], while EPNet achieved the testing error rate of 0.1927 [19.27 percent]. The largest CNN evolved by EPNet among 30 runs had only 6 hidden nodes. The average was 3.4.

Table 3.6 Comparison between EPNet and others (Michie, Spiegelhalter and Taylor 1994) in terms of the average testing error rate on the diabetes problem

Algorithm	**EPNet**	**Logdisc**	**DIPOL92**	**Discrim**
Testing E/R[a]	0.224	0.223	0.224	0.225
Algorithm	**SMART**	**RBF**	**ITrule**	**BP**
Testing E/R[a]	0.232	0.243	0.245	0.248
Algorithm	**Cal5**	**CART**	**CASTLE**	**Quadisc**
Testing E/R[a]	0.250	0.255	0.258	0.262

Note: a E/R indicates the error rate

Table 3.7 shows results on the *heart disease problem* from EPNet and other neural and non-neural algorithms. The GM algorithm (Roy, Govil and Miranda 1995) is used to construct RBF [radial basis function] networks. It produced a RBF network of 24 Gaussians with 18.18 percent testing error. Bennet and Mangasarian (1992) reported a testing error rate of 16.53 percent with their MSM1 method, 25.92 percent with their MSM method, and about 25 percent with BP, which is much worse than the worst CNN evolved by EPNet. The best manually designed CNN achieved 14.78 percent testing error (Prechelt 1994), which is worse than the best result of EPNet, 13.235 percent.

Table 3.7 Comparison between MSM1 (Bennett and Mangasarian 1992), a hand-designed CNN (Prechelt 1994) and EPNet on the heart disease problem. The smallest error rate achieved by EPNet was 0.13235.

	Results MSM1	**Best Results HDCNNS Trial-and-Error**	**Average Results EPNet 30 Runs**
Hidden Nodes	-	4	4.1
Testing E/R	0.1653	0.1478	0.16767

Note: '-' in the table means 'not available'.

Schiffmann, Joost and Werner (1992) tried the *thyroid problem* using a 21:20:3 network. They found that several thousand learning passes were necessary to achieve a testing error rate of 2.6 percent for this network. They also used their genetic algorithm to train multilayer CNNs on the reduced training data set containing 713 examples. They obtained a network with 50 hidden nodes and 278 connections, which had a testing error rate of 2.5 percent. These results are even worse than those generated by the worst CNN evolved by EPNet. However, the best manually designed CNN (Prechelt 1994) has a testing error rate of 1.278 percent, which is better than EPNet's best result, 1.634 percent. This is the only case where the best manually designed CNN (Prechelt 1994) outperforms EPNet's best. The above results are summarized in Table 3.8.

The Australian Credit Card Assessment Problem

One of the things often overlooked in evolutionary algorithms is the information contained in the final population of the evolution. Most researchers just use the best individual in the population without thinking of exploring possible useful information in the rest of the population. In EPNet, simulated evolution is driven by an EP algorithm without any recombination operator. Due to the many-to-many mapping between genotypes and phenotypes (Fogel 1995b), different individuals in the final population may have similar error rates but quite different architectures and weights. Some of them may have overfitted the training and/or validation set, and some may not. In order to avoid overfitting and achieve better generalization, a second validation set has been proposed to stop training in the last step of EPNet.

Table 3.8 Comparison between Schiffmann, Joost and Werner's (1993) best results, a hand-designed CNN (Prechelt 1994) and EPNet on the thyroid problem

	Results Schiffmann	Best Results HDCNNS Trial-and-Error	Average Results EPNet 30 Runs
Hidden Nodes	50	12	5.9
Testing E/R	0.025	0.01278	0.02115

Note: The smallest error rate achieved by EPNet was 0.01634.

In the following experiment, the original validation set was divided into two equal subsets; the first [V-Set 1] was used in the fitness evaluation and the second [V-Set 2] was used in the last step of EPNet. In the last step, all individuals in the final population were first trained by the MBP on the training set and V-Set 1 combined. The one which produced the minimum error rate on V-Set 2 was then chosen as the final output from EPNet and tested on the testing set. Ties were broken in favour of the network with the minimum number of connections. If there was still a tie, it was broken at random.

The effectiveness of using a second validation set in EPNet was tested on another difficult problem – the Australian credit card assessment problem. The problem is to assess applications for credit cards based on a number of attributes. There are 690 cases in total. The output has two classes. The 14 attributes include 6 numeric values and 8 discrete ones, the latter having from 2 to 14 possible values. This data set was also obtained from the UCI machine learning repository. The input attributes used for CNNs are rescaled to between 0.0 and 1.0 by a linear function.

Experimental Results and Comparisons. The whole data set was first randomly partitioned into training data [518 cases] and testing data [172 cases]. The training data was then further partitioned into three subsets: (1) the training set [346 cases]; (2) validation set 1 [86 cases]; and (3) validation set 2 [86 cases].

Table 3.9 Architectures of evolved CNNs for the Australian credit card data set

	Mean	Standard Deviation	Minimum	Maximum
Number of Connections	88.03	24.70	43	127
Number of Hidden Nodes	4.83	1.62	2	7

Table 3.10 Accuracy of evolved CNNs for the Australian credit card data set

	Training Set		V-Set 1		V-Set 2		Test Set	
	E	E/R	E	E/R	E	E/R	E	E/R
Mean	9.69	0.111	6.84	0.074	9.84	0.091	10.17	0.115
Standard Deviation	2.05	0.021	1.47	0.022	2.31	0.024	1.55	0.019
Minimum	6.86	0.081	4.01	0.035	5.83	0.047	7.73	0.081
Maximum	15.63	0.173	9.94	0.105	14.46	0.140	14.08	0.157

Notes: 'V-Set 1' and 'V-Set 2' indicate validation set 1 and validation set 2 respectively, 'E' and 'E/R' indicate error and error rate.

Table 3.11 Comparison between EPNet and others (Michie, Spiegelhalter and Taylor 1994) in terms of the average testing error rate

Algorithm	EPNet	Cal5	ITrule	DIPOL92
Testing E/R	0.115	0.131	0.137	0.141
Algorithm	**Discrim**	**Logdisc**	**CART**	**RBF**
Testing E/R	0.141	0.141	0.145	0.145
Algorithm	**CASTLE**	**NaiveBay**	**IndCART**	**BP**
Testing E/R	0.148	0.151	0.152	0.154

The experiments used the same parameters as those for the diabetes problem, except for the maximum number of generations which was set at 100. The average results over 30 runs are summarized in Table 3.9 and Table 3.10. Very good results have been achieved by EPNet. For example, a CNN with only 2 hidden nodes and 43 connections could achieve an error rate of 10.47 percent on the testing set. Table 3.11 compares EPNet's results with those produced by other algorithms (Michie, Spiegelhalter and Taylor 1994). It is worth pointing out that the other results were obtained by 10-fold cross validation (Michie, Spiegelhalter and Taylor 1994). They represented the best 11 out of 23 algorithms tested (Michie, Spiegelhalter and Taylor 1994). It is clear that EPNet performed much better than others, even though they used 10-fold cross validation.

The MacKey-Glass Chaotic Time Series Prediction Problem

The problem discussed next is different from the previous ones in that its output is continuous. It is not a classification problem, but a function approximation problem for time series prediction. This problem is used to illustrate that EPNet is applicable to a wide range of problems, since it does not assume any *a priori* knowledge of the problem domain. The only part of EPNet which needs changing in order to deal with the continuous output is the fitness evaluation module.

The MacKey-Glass time series investigated here is generated by the following differential equation

$$\dot{x}(t) = \beta\, x(t) + \frac{\alpha\, x(t-\tau)}{1 + x^{10}(t-\tau)} \tag{3.12}$$

where $\alpha = 0.2$, $\beta = -0.1$, $\tau = 17$ (Farmer and Sidorowich 1987; Mackey and Glass 1977). As mentioned by Martinetz, Berkovich and Schulten (1993), $x(t)$ is quasi-periodic and chaotic with a fractal attractor dimension 2:1 for the above parameters.

The input to a CNN consists of four past data points, $x(t)$, $x(t-6)$, $x(t-12)$ and $x(t-18)$, the output is $x(t+6)$. In order to make multiple step prediction (i.e. $\Delta t = 90$) during testing, iterative predictions of $x(t+6)$, $x(t+12)$, ..., $x(t+90)$ will be made. During training, the true value of $x(t+6)$ was used as the target value. Such experimental set-up is the same as that used by Martinetz, Berkovich and Schulten (1993).

In the following experiments, the data for the MacKey-Glass time series was obtained by applying the fourth-order Runge-Kutta method to Equation (3.12) with initial condition $x(0) = 1.2$, $x(t-\tau) = 0$ for $0 \le t < \tau$, and the time step is 1. The training data consisted of point 118 to 617 [i.e. 500 training patterns]. The following 500 data points [starting from point 618] were used as testing data. The values of training and testing data were rescaled linearly to between 0.1 and 0.9. No validation sets were used in the experiments. This experimental set-up was adopted in order to facilitate comparison with other existing work.

The normalized root-mean-square [RMS] error E was used to evaluate the performance of EPNet, which is determined by the RMS value of the absolute prediction error for $\Delta t = 6$, divided by the standard deviation of $x(t)$ (Farmer and Sidorowich 1987, Martinetz, Berkovich and Schulten 1993).

$$E = \frac{\left\langle \left[x_{pred}(t,\Delta t) - x(t+\Delta t)\right]^2 \right\rangle^{\frac{1}{2}}}{\left\langle (x - \langle x \rangle)^2 \right\rangle^{\frac{1}{2}}} \tag{3.13}$$

where $x_{pred}(t; \Delta t)$ is the prediction of $x(t + \Delta t)$ from the current state $x(t)$ and $\langle x \rangle$ represents the expectation of x. As indicated by Farmer and Sidorowich (1987), 'If $E = 0$, the predictions are perfect; $E = 1$ indicates that the performance is no better than a constant predictor $x_{pred}(t; \Delta t) = \langle x \rangle$.'

The following parameters were used in the experiments: the maximum number of generations [200], the number of hidden nodes for each individual in the initial population [8 to 16], the initial learning rate (0.1), the range of learning rate (0.1 to 0.75), the number of mutated hidden nodes (1), the two parameters for training each individual in the initial population [1,000 and 5], and the two parameters for each partial training during evolution [200 and 5]. All other parameters were the same as those for the medical diagnosis problems.

Experimental Results and Comparisons. Table 3.12 shows the average results of EPNet over 30 runs. The error in the table refers to the error defined by Equation (3.6). Table 3.13 compares EPNet's results with those produced by back-propagation training [BP] and cascade-correlation [CC] learning (Crowder 1990). EPNet evolved much more compact CNNs than the cascade-correlation networks, which are more than six times larger than the EPNet-evolved CNNs. EPNet-evolved CNNs also generalize better than the cascade-correlation networks. Compared with the networks produced by BP, the EPNet-evolved CNNs used only 103 connections [the median size] and achieved comparable results.

Table 3.12 The average results produced by EPNet over 30 runs for the MacKey-Glass time-series prediction problem

	Mean	Standard Deviation	Minimum	Maximum
Number of Connections	103.33	24.63	66.00	149.00
Number of Hidden Nodes	10.87	1.78	8.00	14.00
Error on Training Set	0.0188	0.0024	0.0142	0.0237
Error on Testing Set [Δ= 6]	0.0205	0.0028	0.0152	0.0265
Error on Testing Set [Δ = 90]	0.0646	0.0103	0.0487	0.0921

Table 3.13 Comparison of generalization results for the MacKey-Glass time series prediction problem

Method	Number of Connections	Testing Error Δt=6	Testing Error Δt=84
EPNet	103	0.02	0.06
BP	540	0.02	0.05
CC Learning	693	0.06	0.32

For a large time span, $\Delta = 90$, EPNet's results also compare favourably with those produced by Martinetz, Berkovich and Schulten (1993) which have been shown to be better than Moody and Darken (1989). The average number of connections [weights] in an EPNet-evolved CNN is 103.33, while the smallest 'neural-gas' network has about 200 connections [weights] (Martinetz, Berkovich and Schulten 1993), which is almost twice as large as the average size of EPNet-evolved CNN. To achieve a prediction error of 0.05, a neural-gas network had to use 1000 training data points and about 500 connections [weights] (Martinetz, Berkovich and Schulten 1993). The smallest prediction error among 30 EPNet runs was 0.049, while the average prediction error was 0.065. For the same training set size of 500 data points, the smallest prediction error achieved by neural-gas networks was about 0.06. The network achieving the smallest prediction error had 1,800 connections [200 hidden nodes], which is more than 10 times larger than the largest EPNet-evolved CNN.

3.5 Evolutionary Learning and Optimization

Learning is often formulated in the machine learning field as an optimization problem [see Chapter 8 in this volume for discussion]. For example, backpropagation is often used to train feedforward CNNs (Rumelhart, Hinton and Williams 1986). This training process is also called the *learning* process of CNNs. BP is known as one of the most widely used *learning* [training] algorithms. However, BP is in essence a gradient-based optimization algorithm which is used to minimize an error function [often the mean square error] of CNNs. The so-called learning problem here is a typical optimization problem in numerical analysis. Many improvements on the CNN learning algorithm are actually improvements over the optimization algorithms (Hush and Horne 1993), such as conjugate gradient methods (Johansson, Dowla and Goodman 1991, Møller 1993, Fischer and Staufer 1999) and Quasi-Newton-procedures (Fischer and Staufer 1999).

Learning is, in practice, different from optimization because we want the learned system to have best generalization, which is different from minimizing an error function [see Chapter 8 in this volume for reference to the application domain of spatial interaction modelling]. The CNN with the minimum error does not necessarily mean that it has best generalization, unless there is an equivalence between generalization and the error function. Unfortunately, it is almost impossible in practice to measure generalization exactly and accurately (Wolpert 1990), although there are many theories and criteria, such as the minimum description length [MDL] (Rissanen 1978), Akaike information criteria [AIC] (Akaike 1974), and minimum message length [MML] (Wallace and Patrick 1991). In practice, these criteria are often used to define better error functions in the hope that minimizing the functions will maximize generalization. While better error functions often lead to better generalization of learned systems, there is no guarantee. Regardless of the error functions used, BP or other more advanced learning algorithms are still used as *optimization algorithms*. They simply optimize different error functions. The nature of the problem is unchanged.

Similar situations occur with other machine learning methods, where an 'error' function has to be defined. A 'learning' algorithm then tries to *minimize* the function. However, no error functions can guarantee that they correspond to the true generalization (Wolpert 1990). This is a problem faced by most inductive learning methods. There is no way to get around this except by using a good empirical function, which might not correspond to the true generalization. Hence, formulating learning as optimization in this situation is justified.

Evolutionary learning is a population-based learning method. Most scholars use an evolutionary algorithm to maximize a fitness function or minimize an error function, and thus face the same problem as that described above. Maximizing a fitness function is different from maximizing generalization. The evolutionary algorithm is actually used as an optimization, not learning, algorithm. While little can be done for traditional non-population-based learning, there are opportunities for improving population-based learning, e.g., evolutionary learning.

Since the maximum fitness is not equivalent to best generalization in evolutionary learning, the best individual with the maximum fitness in a population may not be the one we want. Other individuals in the population may contain some useful information that will help to improve generalization of learned systems. It is thus beneficial to make use of the whole population rather than any single individual. A population always contains at least as much information as any single individual. Hence, combining different individuals in the last generation to form an integrated system is expected to produce better results.

3.6 A Population of ECNNs as an Ensemble

As discussed in Section 3.5, the previous implementation of EPNet used EP as an optimization algorithm to minimize the CNN's error rate on a validation set. The best individual was always chosen as the final output, and the rest of the population was discarded. However, an individual with the minimum error rate on a validation set might not have the minimum error rate on an unseen testing set. The rest of the population may contain some useful information for improving generalization of ECNNs. In order to integrate useful information in different individuals in the last generation, we can treat each individual as a module, and combine them in a linear fashion (Yao and Liu 1998a). We will call this combined system an 'ensemble' of ECNNs. Linear combination was chosen because of its simplicity. The purpose here is not to find the best combination method, but to show the importance of using population information and the advantage of combining ECNNs. Better results would be expected if we had used non-linear combination methods.

Majority Voting

The simplest linear combination method is majority voting. That is, the output of the most number of ECNNs will be the output of the ensemble. If there is a tie, the output of the ECNN [among those in the tie] with the lowest error rate on V-Set 2 will be selected as the ensemble output. The ensemble in this case is the whole population. All individuals in the last generation participate in voting and are treated equally.

The results of majority voting on the three problems are given in Table 3.14. The majority voting method outperformed the single best individual in two out of three problems. This is rather surprising since majority voting did not consider the differences among the individuals. It performed worse than the best individual on the heart disease problem probably because it treated all individuals in the population equally. However, not all individuals are of equal importance. Some may perform poorly due to a mutation in the previous generation. The greatest advantage of majority voting is its simplicity. It requires virtually no extra computational cost.

Table 3.14 Accuracies of the ensemble formed by majority voting [results averaged over 30 runs]

		Error Rate			
		Train	**V-Set 1**	**V-Set 2**	**Test**
Credit Card Assessment Problem	Mean	0.128	0.048	0.133	0.095
	Standard Deviation	0.013	0.015	0.016	0.012
	Minimum	0.101	0.023	0.105	0.076
	Maximum	0.150	0.081	0.151	0.122
Diabetes Problem	Mean	0.218	0.175	0.169	0.222
	Standard Deviation	0.020	0.024	0.020	0.022
	Minimum	0.177	0.135	0.115	0.172
	Maximum	0.276	0.240	0.208	0.255
Heart Disease Problem	Mean	0.083	0.002	0.057	0.167
	Standard Deviation	0.025	0.011	0.017	0.024
	Minimum	0.052	0.000	0.029	0.132
	Maximum	0.157	0.059	0.088	0.235

Notes: V-Set 1 indicates validation set 1 and V-Set 2 validation set 2.

Table 3.15 shows the results of the t-test comparing the best individual to the ensemble formed by majority voting. At the 0.05 level of significance, the ensemble is better than the best individual for the Australian credit card and diabetes problems and worse for the heart disease problem.

Table 3.15 t-test values comparing the best individual to the ensemble formed by majority voting [values calculated on basis of 30 runs]

	t-Test Value		
	Credit Card Assessment Problem	**Diabetes Problem**	**Heart Disease Problem**
Ensemble - Individual	2.5362	−1.8024	−2.7186

Rank-Based Linear Combination

One way to consider differences among individuals without entailing too much extra computational cost is to use the fitness information to compute a weight for each individual. In particular, we can use ECNN ranking to generate weights for each ECNN in combining the ensemble output. That is, given N sorted ECNNs with an increasing error rate on V-Set 2, where N is the population size, and their outputs $o_1, \ldots, o_N$, then the weight for the ith ECNN is

$$w_i = \frac{\exp(\beta(N+1-i))}{\sum_{j=1}^{N} \exp(\beta_j)} \tag{3.14}$$

where β is a scaling factor. It was set as 0.75, 0.5, and 0.75 for the Australian credit card, diabetes and heart disease problems, respectively. These numbers were selected after modest preliminary experiments with $\beta = 0.1$; 0,25; 0.5 and 0.75. The ensemble output is

$$O = \sum_{j=1}^{N} w_j o_j \tag{3.15}$$

Table 3.16 Accuracies of the ensemble formed by the rank-based linear combination method [results averaged over 30 runs]

		Error Rate			
		Train	**V-Set 1**	**V-Set 2**	**Test**
Credit Card Assessment Problem	Mean	0.112	0.063	0.123	0.095
	Standard Deviation	0.017	0.019	0.014	0.012
	Minimum	0.081	0.035	0.093	0.070
	Maximum	0.156	0.105	0.151	0.116
Diabetes Problem	Mean	0.213	0.185	0.171	0.225
	Standard Deviation	0.020	0.030	0.021	0.023
	Minimum	0.172	0.125	0.125	0.172
	Maximum	0.245	0.260	0.198	0.271
Heart Disease Problem	Mean	0.057	0.003	0.044	0.154
	Standard Deviation	0.023	0.009	0.022	0.031
	Minimum	0.007	0.000	0.000	0.088
	Maximum	0.119	0.029	0.088	0.235

Note: V-Set 1 indicates validation set 1 and V-Set 2 validation set 2.

The results of the rank-based linear combination method are given in Table 3.16. In this case, the results produced by the ensemble are either better than or as good as those produced by the best individual. Table 3.17 shows the results of the t-test comparing the best individual to the ensemble formed by the rank-based linear combination. The ensemble is better than the best individual for the Australian credit card and diabetes problems at the 0.05 level of significance. It also outperforms the best individual for the heart disease problem [though there is no statistical significance].

Table 3.17 t-test values comparing the best individual to the ensemble formed by rank-based linear combination [values calculated on basis of 30 runs]

	t-Test Value		
	Credit Card Assessment Problem	**Diabetes Problem**	**Heart Disease Problem**
Ensemble - Individual	−2.3000	−3.4077	−0.1708

Linear Combination by the RLS Algorithm

One of the well-known algorithms for learning linear combination weights [i.e. one-layer linear networks] is the recursive least square [RLS] algorithm (Mulgrew and Cowan 1988, pp. 31-33). It is used to find the weights that minimize the mean square error

$$E = \sum_{i=1}^{M} \left(d(i) - \sum_{j=1}^{N} w_j o_j(i) \right)^2 \tag{3.16}$$

where M is the number of training examples and $d(i)$ is the desired output for example i. [There should be another summation over all outputs of the ensemble on the right-hand side of Equation (3.16). We have omitted it here for the sake of convenience.] Minimizing the error E with respect to weight w_k yields

$$\frac{\partial E}{\partial w_k} = -2\sum_{i=1}^{M} \left(d(i) - \sum_{j=1}^{N} w_j o_j(i) \right) o_k(i) = 0 \tag{3.17}$$

Equation (3.17) can be expressed in matrix form

$$\boldsymbol{r}_{oo}\, \boldsymbol{w} = \boldsymbol{r}_{od} \tag{3.18}$$

where

$$\boldsymbol{r}_{oo} = \sum_{i=1}^{M} \boldsymbol{o}(i)\ \boldsymbol{o}^T(i) \tag{3.19}$$

and

$$\boldsymbol{r}_{od} = \sum_{i=1}^{M} d(i)\ \boldsymbol{o}(i) \tag{3.20}$$

A unique solution to Equation (3.18) exists if the correlation matrix $\boldsymbol{r}_{oo}$ is non-singular. The weight vector w could be found by inverting $\boldsymbol{r}_{oo}$ and multiplying it by $\boldsymbol{r}_{od}$ according to Equation (3.18). However, this method is time-consuming because whenever a new training example becomes available it requires inversion and multiplication of matrices. The RLS algorithm (Mulgrew and Cowan 1988, pp. 31-33) uses a different method to determine the weights. From Equation (3.19) and Equation (3.20) we can get

$$\boldsymbol{r}_{oo}(i) = \boldsymbol{r}_{oo}(i-1) + \boldsymbol{o}(i)\ \boldsymbol{o}^T(i) \tag{3.21}$$

and

$$\boldsymbol{r}_{od}(i) = \boldsymbol{r}_{od}(i-1) + d(i)\ \boldsymbol{o}(i) \tag{3.22}$$

Using Equations (3.18), (3.21) and (3.22), we can get

$$\boldsymbol{w}(i) = \boldsymbol{w}(i-1) + \boldsymbol{k}(i)\ e(i) \tag{3.23}$$

where

$$e(i) = d(i) - \boldsymbol{w}^T(i-1)\ \boldsymbol{o}(i) \tag{3.24}$$

and

$$\boldsymbol{k}(i) = \boldsymbol{r}^{-1}_{oo}(i)\ \boldsymbol{o}(i) \tag{3.25}$$

A recursion for $r^{-1}_{oo}(i)$ is given by (Mulgrew and Cowan 1988)

$$r_{oo}^{-1}(i) = r_{oo}^{-1}(i-1) - \frac{r_{oo}^{-1}(i-1)\ o(i)\ o^{T}(i)\ r_{oo}^{-1}(i-1)}{1 + o^{T}(i)\ r_{oo}^{-1}(i-1)\ o(i)} \tag{3.26}$$

In our implementation of the above RLS algorithm, three runs were always performed with different initial $r^{-1}_{oo}(0)$ and weights $w(0)$. The initial weights were generated at random in [−0.1, 0.1].

$$r^{-1}_{oo}(0) = \alpha\ I_N \tag{3.27}$$

where $\alpha = 0.1$, 0.2 and 0.3, and I_N is an N x N unit matrix. The best result out of the three runs was chosen as the output from the RLS algorithm. The results of the ensemble formed by the RLS algorithm are given in Table 3.18. It is clear that the ensemble performed better than the best individual for all three problems. The results also indicate that a better combination method can produce better ensembles. In fact, the RLS algorithm is one of the recommended algorithms for performing linear combinations (Hashem 1993; Perrone 1993). However, other algorithms (see Baldi and Hornik 1995) can also be used.

Table 3.18 Accuracies of the ensemble formed by the RLS algorithm [results averaged over 30 runs]

		Error Rate			
		Train	**V-Set 1**	**V-Set 2**	**Test**
Credit Card Assessment Problem	Mean	0.114	0.042	0.125	0.093
	Standard Deviation	0.018	0.012	0.014	0.011
	Minimum	0.087	0.023	0.093	0.076
	Maximum	0.147	0.070	0.163	0.116
Diabetes Problem	Mean	0.211	0.159	0.171	0.226
	Standard Deviation	0.024	0.025	0.017	0.021
	Minimum	0.169	0.104	0.115	0.193
	Maximum	0.255	0.208	0.208	0.260
Heart Disease Problem	Mean	0.058	0.000	0.039	0.151
	Standard Deviation	0.027	0.000	0.021	0.033
	Minimum	0.015	0.000	0.000	0.088
	Maximum	0.119	0.000	0.088	0.221

Table 3.19 shows the results of the t-test comparing the best individual to the ensemble formed by the RLS algorithm. The ensemble is better than the best individual at the 0.05 level of significance for the Australian credit card and diabetes problems, and better at the 0.25 level of significance for the heart disease problem.

Table 3.19 t-test values comparing the best individual to the ensemble formed by the RLS algorithm [values calculated on basis of 30 runs]

	t-TestValue		
	Credit Card Assessment Problem	Diabetes Problem	Heart Disease Problem
Ensemble - Individual	−2.7882	−1.9046	−0.7862

Using a Subset of the Population as an Ensemble

For the previous three combination methods, all the individuals in the last generation were used in the ensembles. It is interesting to investigate whether we can reduce the size of the ensembles without increasing the testing error rates too much. Such an investigation can provide some hints on whether all the individuals in the last generation contain some useful information, and also shed some light on the importance of a population in evolutionary learning.

As the space of possible subsets is very large [$2^N - 1$] for a population of size N, it is impractical to use an exhaustive search to find an optimal subset. Instead, we used a genetic algorithm [GA] (Goldberg 1989) to search for a near optimal subset. The weights for each ECNN in each subset were determined by the same RLS algorithm described above. The GA used the following parameters: population size [50], maximum number of generations [50], crossover rate [0.6], mutation rate [0.01], two-point crossover and bit-string length [20]. These parameters were chosen somewhat arbitrarily, and are not necessarily the best. Elitism was used in the GA, of which the major steps are summarized in Fig. 3.6.

(ix) Initialize the population at random.
(x) Train each individual by the RLS algorithm and use the result to define the individual's fitness.
(xi) Reproduce a number of children for each individual in the current generation based on a non-linear ranking scheme (Yao 1993a).
(xii) Apply crossover and mutation to the children generated above and obtain the next generation.

Fig. 3.6 Major steps of the evolution of ensemble structures

The results of the ensemble formed by a subset of the last generation are given in Table 3.20. The same GA described above was used to search for near optimal subsets for all three problems. Table 3.21 gives the sizes of the subsets evolved. It is clear that sizes have been reduced on average by 50 percent. Table 3.22 shows t-test values comparing the accuracies of the best individual to those of the ensemble. For the Australian credit card and diabetes problems, the ensemble is better than the best individual at the 0.10 and 0.005 level of significance, respectively. It is worse than the best individual for the heart disease problem at the 0.05 level of significance. This result might be caused by the small number of generations [only 50] used in our experiments. A larger number could probably produce better results, but would increase the search time. Letting the GA run for 100 generations for the heart disease problem, the average testing error rate over 30 runs improved from 0.164 to 0.159. A t-test revealed that the ensemble was only worse than the best individual at the 0.10 level of significance for the heart disease problem.

Table 3.20 Accuracies of the ensemble formed by a near optimal subset of the last generation [results averaged over 30 runs]

		Error Rate			
		Train	**V-Set 1**	**V-Set 2**	**Test**
Credit Card Assessment Problem	Mean	0.117	0.036	0.106	0.095
	Standard Deviation	0.018	0.007	0.012	0.012
	Minimum	0.087	0.023	0.081	0.070
	Maximum	0.159	0.058	0.128	0.116
Diabetes Problem	Mean	0.219	0.129	0.160	0.222
	Standard Deviation	0.025	0.024	0.016	0.023
	Minimum	0.174	0.094	0.125	0.182
	Maximum	0.268	0.167	0.188	0.260
Heart Disease Problem	Mean	0.068	0.000	0.017	0.164
	Standard Deviation	0.028	0.000	0.017	0.030
	Minimum	0.022	0.000	0.000	0.118
	Maximum	0.134	0.000	0.059	0.221

Note: V-Set 1 indicates validation set 1 and V-Set 2 validation set 2.

Table 3.21 Ensemble size found by the GA [results averaged over 30 runs]

	Credit Card Assessment Problem	**Diabetes Problem**	**Heart Disease Problem**
Mean	10.50	8.40	10.30
Standard Deviation	3.78	2.14	4.56
Minimum	3.00	5.00	1.00
Maximum	20.00	12.00	19.00

Table 3.22 t-test values comparing the best individual to the ensemble formed by a subset of the last generation [values calculated on basis of 30 runs]

		t-Test Value	
	Credit Card Assessment Problem	**Diabetes Problem**	**Heart Disease Problem**
Ensemble - Individual	−1.6247	−2.7904	1.7250

3.7 Conclusions

This chapter has described an evolutionary system, EPNet, for designing and training CNNs. The idea behind EPNet is to put more emphasis on evolving CNN behaviours, rather than its circuitry. A number of techniques have been adopted in EPNet to maintain a close behavioural link between parents and their offspring. For example, partial training is always employed after each architectural mutation in order to reduce the behavioural disruption to an individual. The training mutation is always attempted first, before any architectural mutation, since it causes less behavioural disruption. A hidden node is not added to an existing CNN at random, but through splitting an existing node.

It was observed in our experiments that the performance of evolved CNNs deteriorated when the number of epochs and steps in partial training was lower than certain values. That is, a weak behavioural link between the parent and its offspring will have a negative impact on the system's performance.

In order to reduce the noise in fitness evaluation, EPNet evolves CNN architectures and weights simultaneously. Each individual in a population evolved by EPNet is a CNN with weights. The evolution simulated by EPNet is closer to the Lamarckian than Darwinian evolution. Weights and architectures learned by one generation are inherited by the next generation. This is quite different from most genetic approaches, where only architectures and not weights are passed to the next generation.

EPNet encourages parsimony of evolved CNNs by ordering their mutations rather than using a complexity [regularization] term in the fitness function. It avoids the tedious trial-and-error process in determining the coefficient for the complexity term. The effectiveness of the method has been shown by the compact CNNs evolved by EPNet, which have very good generalization ability.

EPNet has been tested on a number of benchmark problems, including the N parity problem, the two-spiral problem, as well as the medical diagnosis, Australian credit card assessment, and MacKey-Glass time series prediction problems. Very competitive results have been produced by EPNet in comparison with other algorithms. EPNet imposes very few constraints on feasible CNN architectures, and thus faces a huge search space of different CNNs. It can escape from structural local minima due to its global search capability. The experimental results have shown that EPNet can explore the CNN space effectively. It can discover novel CNNs which would be very difficult for human beings to design.

It is emphasized in this chapter that although learning problems are often formulated as optimization ones, learning is very different from optimization in practice. Population-based learning, and evolutionary learning in particular, should exploit this difference by making use of the information in a population. It was demonstrated that a population contains more information than any individual within it. Such information can be used effectively to improve generalization of the learning systems.

The four methods introduced here are all linear combinations. The first two involve little extra computational cost. In particular, the rank-based combination method makes use of the fitness information readily available from EPNet to compute weights and achieves good results. This method fits evolutionary learning well due to its simplicity and effectiveness. The other two combination methods are based on the well-known RLS algorithm. They require a little more computational time, but produce good results. If computation time is not of primary concern, other linear (Baldi and Hornik 1995) or non-linear methods can also be used.

This chapter shows one way of evolving a CNN ensemble, where individuals try to solve a complex problem collectively and cooperatively. However, no special considerations were made in the evolution of CNNs to encourage such cooperation. One of our future projects will be to encourage the evolution of cooperating individuals in a population through techniques such as fitness sharing (Darwen and Yao 1995, 1996b, 1997) and negative correlation (Liu and Yao 1997, 1999a, 1999b).

4 Neural and Evolutionary Computation Methods for Spatial Classification and Knowledge Acquisition

Yee Leung
Department of Geography, The Chinese University of Hongkong

4.1 Introduction

Non-linearity, complexity and dynamics have become a focal point of research in spatial analysis, especially in the analysis of spatial data. Regardless of what we are dealing with, the need to handle systems with a high degree of complexity is now the rule rather than the exception. With our spatial systems becoming more and more complex and highly fluid, non-linearity prevails and evolution is full of surprises. It is essential to develop approaches which can effectively analyse complexity, non-linearity and dynamics in spatial systems in general, and in particular spatial classification and knowledge acquisition.

One of the most fundamental issues in spatial analysis is classification and pattern identification. No matter whether we are dealing with remotely sensed images, digital geographic information [as GIS data] or raw data, we have somehow to classify data into patterns or images into types. The task of classification is to group high dimensional data into separate clusters which represent distinguishable spatial features or patterns.

Over the years, a large variety of techniques have been developed and applied to spatial data classification. Statistical methods are perhaps the most dominant methodology. Discriminant ânalysis, cluster analysis, principal component analysis, factor analysis and other statistical techniques, though with different assumptions and theoretical constructs, have been employed to classify spatial data. If used correctly, they are powerful and effective. One may then wonder why neural and evolutionary computations are necessary when existing statistical methods appear to be handling quite adequately the task of high dimensional data classification. The answer to the question is essentially threefold.

First, classification problems in general have a high degree of non-linearity. Mathematically speaking, data classification is basically a partitioning problem in high dimensional space. We need to find hypersurfaces to separate data into clusters.

For linearly separable problems, statistical methods are usually effective and efficient. These methods fail, however, when non-linearity exists in spatial classification. Spatial data are ordinarily not linearly separable. For example, while linear separability implies spherical separability, and spherical separability implies quadratic separability, the reverse is not true. To be versatile, classification methods therefore need to be able to handle non-linearity. Neural networks, especially multilayer feedforward neural networks, are appropriate for such a task. Learning algorithms in most neural networks can be viewed as a problem of hypersurface reconstruction which tries to approximate the hypersurfaces which partition non-linearly separable clusters from training examples.

Second, most statistical methods assume certain types of probability distributions, such as the Gaussian distribution in the maximum-likelihood classifier. Nevertheless, many spatial data are non-Gaussian. In the classification of remotely sensed images, for example, data are multisource. Gaussian distribution is generally a wrong assumption when spectral data [colour, tone] and spatial data [shape, size, shadow, pattern, texture, and temporal association] are simultaneously employed as a basis of classification (see, for example, Benediktsson, Swain and Esroy 1990; Hepner et al. 1990). Thus conventional methods are not effective for handling scale and aggregation dependence, or a mixture of data types and sources. Neural models (Arbib 1995) and evolutionary models (Bäck, Fogel and Michalewicz 1997), on the other hand, if correctly applied, can give a better solution to the problem.

Third, on-line and real-time computing is a desirable feature of a classifier. Most statistical methods are neither on-line nor real-time. By contrast, neural-network and evolutionary algorithms usually strive for on-line or real-time computation. This is especially important when we are classifying a large volume of data and do not wish to re-classify the data with new data, for example, but still would like to train the classifier with that additional piece of information through a single computation. The adaptive mechanism of some neural-network learning algorithms can perform on-line or real-time computation.

The insurgent interest in neural and evolutionary computation is therefore a response to large volume and variety of spatial data which has to be dealt with nowadays. Digital data and remotely sensed images come from various sources and in various types, resulting in complex data structures with a high degree of non-linearity. Neural and evolutionary computational models can theoretically handle what conventional classification methods fail to manage: multitype and noisy data, uneven reliability in data sources, non-smooth [e.g. non-Gaussian] information distributions, non-linearity, and on-line or real-time computation.

While the neural and evolutionary approaches to spatial classification are gaining momentum, a parallel movement is also taking place in the field of knowledge acquisition. In recent years, our ability to acquire knowledge from examples/data has proved crucial in the construction of intelligent spatial decision support systems for decision-making in highly complex environment (Leung 1997a). Acquiring spatial knowledge, largely captured as rules, is a difficult but important task in this area of research. The process is embedded with non-linearity, complexity and dynamics which are difficult to handle with conventional methods. Neural and evolutionary models, on the other hand, offer promising alternatives. In a way, rule extraction can be viewed as a special type of data clustering/classification in high dimensional space. It thus brings these two major classes of spatial problems, i.e. spatial classification and knowledge acquisition, within the neural and evolutionary framework.

The purpose of this chapter is to provide an analysis of the state-of-the-art of developments and applications of neural and evolutionary computations in spatial classification and knowledge acquisition. In Sections 4.2, 4.3 and 4.4 a number of applications of neural and evolutionary computations in spatial classification and pattern analysis are examined. Their effectiveness is scrutinised and some problems in their use also raised. In Section 4.5, neural-based scale-space algorithms are introduced for spatial classification by clustering. For rule acquisition, a radial basis function neural network and a hybrid fuzzy neural network are analyzed in Sections 4.6 and 4.7 respectively. To show how evolutionary computations can be applied to acquire precise and fuzzy rules, two genetic-algorithm-based systems: SCION and GANGO are then introduced in Sections 4.8 and 4.9. The chapter concludes with a summary and outlook for further research.

4.2 Spatial Classification by Multilayer Feedforward Neural Networks

Among the various neural network models, multilayer feedforward neural networks are perhaps one of the most powerful for spatial data classification, especially when non-linearity is encountered in the data set (Leung 1997b). A special case of multilayer networks is the single-layer perceptron which is capable of handling linear separability.

A single-layer perceptron comprises an input array and an output layer. Let $\boldsymbol{X} = \{x_1, \ldots, x_m\}$ and $\boldsymbol{Y} = \{y_1, \ldots, y_n\}$ be the input and output vectors respectively. In the limiting case, there can be only one node in the output layer (Fig. 4.1). Given a set of input-output patterns, an initial weight matrix $\boldsymbol{w} = \{w_{ij}\}$ and a threshold vector θ, the output y_j of the jth neuron can be obtained as

$$y_i = \mathrm{F}\left(\sum_{i=1}^{m} w_{ij}\, x_i - \theta_j\right) \quad j = 1, \ldots, n \tag{4.1}$$

where $\mathrm{F}(\cdot)$ is the activation [transfer] function which can be discrete, i.e. a step-function, or continuous, i.e. a sigmoid function; θ_j is a specified threshold value attached to node j; and w_{ij} is the connection weight from input i to output j. A simple encoding scheme is a sign function

$$y_i = F\left(\sum_{i=1}^{m} w_{ij}\, x_i - \theta_j\right) = \begin{cases} 1 & \text{if} \sum_{i=1}^{m} w_{ij}\, x_i - \theta_j \geq 0 \\ 0 & \text{otherwise}. \end{cases} \tag{4.2}$$

To put it in the perspective of spatial data classification, the input array ***X*** encodes relevant data, e.g. spectral reflectance in various bands, and the output layer ***Y*** encodes spatial patterns, e.g. land types. A learning algorithm is used to learn the internal representation of the data, i.e. associating appropriate input and output patterns, by adaptively modifying the connection weights of the network. For the single-layer perceptron, given the encoding scheme in (4.2), the initial weights can be recursively adjusted by

$$w_{ij}(t+1) = w_{ij}(t) + \Delta w_{ij} \tag{4.3}$$

where $w_{ij}(t)$ is the connection weight from input i to neuron j at time t; and Δw_{ij} is the change in weight which may be obtained by the delta rule:

$$\Delta w_{ij} = \eta\, \delta_j\, x_i \tag{4.4}$$

where η, $0 < \eta < 1$, is a learning rate [step size], and δ_j is the error at processing unit j obtained as

$$d_j = y_j^e - y_j \tag{4.5}$$

where y_j^e and y_j are respectively the desirable value and the output value of processing unit j. The iterative adjustment terminates when the process converges with respect to some stopping criteria such as a specified admissible error.

A drawback of the single-layer perceptron is that it cannot handle non-linearity. The famous XOR problem is a typical example where the perceptron fails. For highly non-linear data, we need to embed hidden layers within the perceptron framework and form the multilayer feedforward architecture (see Fig. 4.2). For example, nodes in the first layer help to create a hyperplane to form separation regions. Nodes in the second layer combine hyperplanes to form convex separation regions. More complex separation regions can be created by adding more hidden layers to the network.

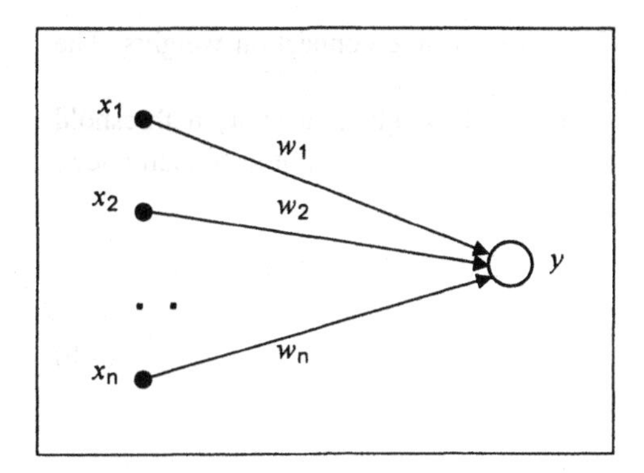

Fig. 4.1 A single-layer perceptron

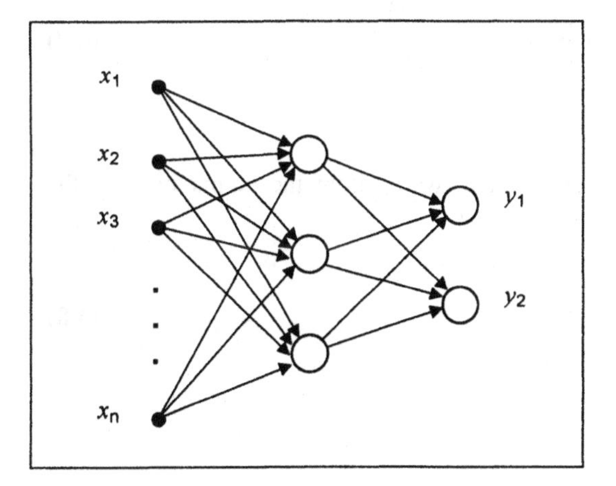

Fig. 4.2 A multilayer feedforward neural network

One of the advantages of using neural networks for spatial data classification is that the system can learn about its environment by dynamically adjusting its connection weights, thresholds, and topology. Mathematically speaking, learning is a problem of hypersurface reconstruction. It determines whether the internal representation to be learnt is separable linearly, spherically, or quadratically. Learning through the iterative adjustment of weights constitutes a good part of the research on learning algorithms.

For multilayer feedforward neural networks, a more powerful learning rule, called gradient descent rule can be used, in combination with the backpropagation technique (Rumelhart and McClelland 1986), to recursively adjust the connection weights. The method is essentially a gradient descent procedure which searches for the solution of an error-minimisation problem along the steepest descent, i.e.

negative gradient, of the error surface with respect to the connection weights. The activation function is a sigmoid function.

Given a set of input-output patterns, an initial weight matrix $\boldsymbol{w}$, a threshold vector θ and a sigmoid activation function, the value y_j of an output or hidden node j is obtained as

$$y_i = \mathrm{F}\left(\sum_{i=1}^{m} w_{ij}\, x_i - \theta_j\right) \tag{4.6}$$

where

$$\mathrm{F}\left(\sum_{i=1}^{m} w_{ij}\, x_i - \theta_j\right) = \left(1 + \exp\left(-\left(\sum_{i=1}^{m} w_{ij}\, x_i - \theta_j\right)\right)\right)^{-1} \tag{4.7}$$

Weight adjustment works recursively backwards from the output node to the hidden layers by

$$w_{ij}(t+1) = w_{ij}(t) + \Delta w_{ij} \tag{4.8}$$

where

$$\Delta w_{ij} = \eta \delta_j x_i \tag{4.9}$$

and η is a learning rate. The error gradient δ_j at the output node j is obtained as

$$\delta_j = y_j(1 - y_j)(y_j^e - y_j) \tag{4.10}$$

and at the hidden node as

$$\delta_j = y_j(1 - y_j) \sum_{k=1}^{\ell} \delta_k w_{jk} \tag{4.11}$$

where δ_k is the error gradient at node k with a connection coming from hidden node j. The iterative procedure stops when convergence is achieved with respect to a stopping criterion.

Remark 4.1 Compared with the maximum-likelihood Gaussian classifier, the advantages of the multilayer feedforward neural network are that:

- The learning algorithm is distribution-free. No critical assumptions are made on the underlying distribution of the spatial data. It is thus more robust in handling non-Gaussian or asymmetrically distributed data. It serves as a model-free approximator of input-output mappings.
- The learning algorithm works directly on the errors without having to know in what form they appear. The model is thus more fault tolerant.
- The learning algorithm is adaptive to real-time changes in weights. Though we can make a maximum-likelihood Gaussian classifier adaptive to changing environment, it takes a great deal of effort and more complex implementation and computation.

Feedforward neural networks have been constructed to process spatial information involving the recognition and classification of remotely sensed images (Fig. 4.3) (Kanellopoulos et al. 1997; see also Chapter 6 in this volume). Multilayer feedforward neural networks with backpropagation have been applied for example to the classification of satellite images (Short 1991; Zhuang et al. 1991; Tzeng et al. 1994; Foody, McCulloch and William 1995; Moddy, Gopal and Strahler 1996; Kimes et al. 1998), to map land cover (Campbell, Hill and Cromp 1989: Hepner et al. 1990; Bischof, Schneider and Pinz 1992; Heerman and Khazenie 1992; Kanellopoulos et al. 1992; Civco 1993; Yoshida and Omatu 1994; Wang 1994; Chen et al. 1995; Fischer et al. 1997; Paola and Schowengerdt 1997) and forest (Wilkinson et al. 1995; Gopal and Woodcock 1996; Skidmore et al. 1997) and to classify cloud types (Lee et al. 1990; Welch et al. 1992; Bankert 1994; Yahn and Simpson 1995; Lewis, Côté and Tatnall 1997; Walder and MacLaren 2000).

Though there are slight variations in topologies, these network models generally consist of one or more hidden layers. The input layer comprises input units through which spectral information of selected channels and/or non-spectral information is supplied via pixel-based and/or region-based schemes [e.g. $n \times n$ window around a pixel]. The output layer consists of land types or imagery types of a classification scheme. The hidden layer contains hidden nodes whose number can be dynamically or manually optimized. Classification is then accomplished by the backpropagation technique in combination with the gradient descent learning rule.

Compared to statistical approaches, neural network models have been shown to be more powerful in the classification of multisource data consisting of spectral properties such as tone or colour, and spatial properties such as shape, size, shadow, pattern, texture, and their associations across a temporal profile (Argialas and Harlow 1990; Benediktsson, Swain and Esroy 1990; Hepner et al. 1990; Civco 1993; Peddle et al. 1994). In general, they can handle what statistical methods, such as the maximum-likelihood Gaussian classifier, fail to manage, i.e. multitype data, unequal reliability of data sources, and non-Gaussian information distributions. To gain the best of both worlds, attempts have been made to integrate neural and statistical approaches to spatial data classification (Wilkinson et al. 1995).

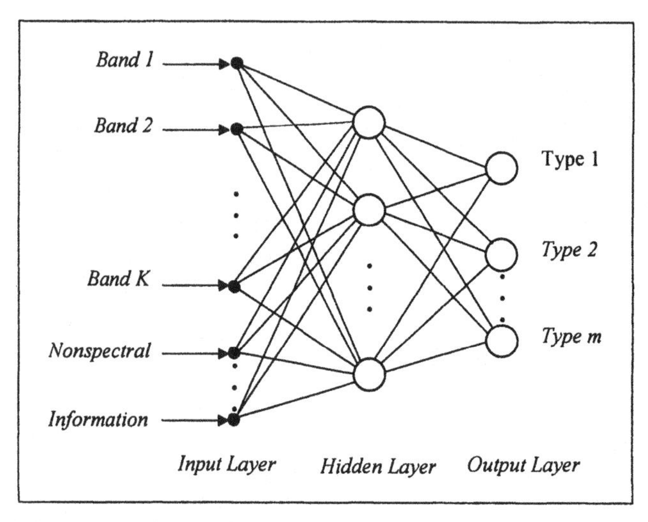

Fig. 4.3 Multilayer feedforward neural network for classification of remotely sensed data

The advantage of backpropagation is that it can handle non-linearity, while a weakness is that the training procedure is prone to local minima. The determination of the right topology can also require a long trial-and-error manual adjustment process. Other disadvantages of the neural network approach to spatial classification and pattern analysis, are the high computation overheads, the risk of overfitting and undertraining (Bruzzone 1997; Skidmore et al. 1997; Wilkinson 1997; Benediktsson and Kanellopoulos 1999). To alleviate the problem, they can be combined with methods such as genetic algorithms (Holland 1975; Goldberg 1989; Davis 1991) to construct network topology and fine tune the weights (see for example Leung et al. 1995; Fischer and Leung 1998, Yao 1999).

4.3 Spatial Classification by Other Unidirectional Neural Networks

Though multilayer feedforward neural networks are able to handle non-linearity, they are not specifically built to handle input noise in data classification. On the other hand, linear associative memories, which are single-layer feedforward neural networks, are constructed for this specific purpose. Such a model can be useful for classifying noisy data in GIS and remote sensing systems for example.

A linear associative memory [LAM] is a single-layer feedforward neural network which stores a given set of pattern pairs in the form of weighted connections between neuron-like elements (Kohonen 1988). This type of network is not only the simplest model of associative memories, but has also had important applications in pattern recognition (Wechsler and Zimmerman 1988) and signal deconvolution (Eichmann and Stojancic 1987).

A LAM is expected to store a set of r pattern pairs $\{X^{(i)}, Y^{(i)}\}$, $X^{(i)} \in \Re^n$, $Y^{(i)} \in \Re^m$, $i = 1, \ldots, r$, in such a way that whenever one of the stored vectors, $X^{(i)}$, is presented with noise, the LAM responds by producing the vector that is closest to the related vector, $Y^{(i)}$, as output (see Fig. 4.4). If $w = (w_{ij})_{n\times m}$ denotes the matrix of connection weights [also referred to as the encoding matrix or connection matrix] between the input and output of the LAM, where w_{ij} represents the connection strength from input unit i to output unit j in the output layer, it is then required that whenever a vector of the form $\tilde{X}_i = X^{(i)} + \varepsilon_i$ is presented as an input, where ε_i is an error or perturbation vector, the output of the LAM

$$\tilde{Y}_i = w\,\tilde{X}_i \tag{4.12}$$

should be as close to $Y^{(i)}$ as possible. Leung, Dong and Xu (1998) have performed a theoretical analysis of the optimal encoding of LAM under the white-noise and coloured-noise circumstances.

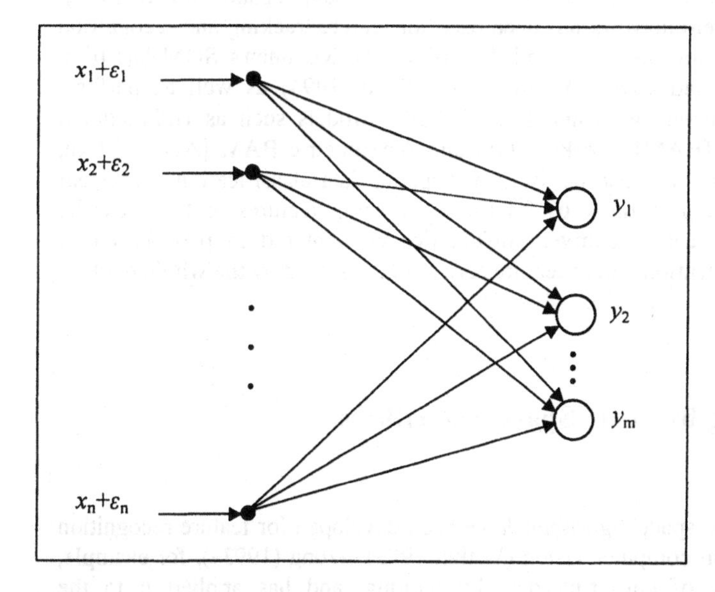

Fig. 4.4 A linear associative memory

In addition to the above two classes of neural networks, there is another class called radial basis function [RBF] neural networks which although powerful are seldom utilized as models for spatial data classification (Fischer et al. 1997). The RBF network is a feedforward neural network with a single hidden layer [the kernel layer]. A set of basis functions, usually non-linear, is introduced into the network, each corresponding to one data point in the input array. Outputs in the output layer are taken to be a linear combination of the basic functions. The hidden layer can be derived by various techniques. Fischer et al. (1997) and Rollet et al. (1998), for example, use the K-means algorithm, while Leung, Lou and Zhou (2000) have used fuzzy ART to obtain the kernel layer in their classifications of remotely sensed data. A competitive mechanism can also be employed to handle the hidden layer in the use of RBF network for rule learning (Leung and Lin 1996) (see Section 4.6 for details).

4.4 Spatial Classification by Recurrent Neural Networks

In addition to the use of unidirectional neural networks for spatial classification and pattern analysis, recurrent networks have also been employed, albeit to a much lesser extent. For example, modified ART (Baraldi and Parmiggiani 1995) and fuzzy ARTMAP (Carpenter et al. 1997; Gopal and Fischer 1997; Mannan, Roy and Ray 1998) have been used to classify remotely sensed images. An ART-like neural network for convex hull computation, the basic building block for high dimensional classification, has also been developed (Leung, Zhang and Xu 1997). Likewise, the Hopfield network has been used for feature tracking and recognition of remotely sensed images (Côté and Tatnall 1997). Kohonen's SOM has been employed to map land cover (Yoshida and Omatu 1994) as well as perform multisource data fusion (Wan and Fraser 1999). Models such as Bidirectional Associate Memory [BAM] (Kosko 1988) and Asymmetric BAM [ABAM] (Xu, Leung and He 1994) may also be useful for spatial data classification. Recurrent neural networks are potentially useful for recognizing features or types by the whole entity. They can circumvent difficulties encountered in pixel-based or region-based classification where features are not recognized by the whole entity.

4.5 Clustering by Scale-Space Algorithms

In recent years, scale-space algorithms have been developed for feature recognition and scene analysis in computer vision (Witkin 1983). Wong (1993a), for example, proposes a concept of clustering data by melting, and has applied it to the classification of multispectral and polarimetric images (Wong 1993b). The algorithm works by melting the system to produce a tree of clusters in the scale space. It has been demonstrated that the classification map is less noisy and more accurate than those obtained by hierarchical rules.

In spatial pattern recognition and image processing, human eyes seem to possess a singular aptitude for grouping objects and finding important structures in an efficient and effective way. Thus a clustering algorithm simulating the visual system may solve some basic problems in these areas of research. Based on scale space theory, a new approach to the mining of clusters in large spatial databases by modelling the blurring effect of lateral retinal interconnections has been formulated. It has been used to classify remotely sensed images (Leung, Zhang and Xu 2000). In this approach, a spatial data set is considered as an image with each light point located at a datum position. As we blur this image, smaller light blobs merge into larger ones until the whole image becomes one light blob at a low enough level of resolution. By identifying each blob with a cluster, the blurring process generates a family of clusterings along the hierarchy. The advantages of this approach are manifold. *First*, the derived algorithms are computationally stable and insensitive to initialization, and they are also totally free from difficult global optimization problems. *Second*, the approach facilitates the construction of new checks on cluster validity and provides the final clustering with a significant degree of robustness to noise in data and change in scale. *Third*, it is more robust in cases where hyper-ellipsoidal partitions may not be assumed. *Fourth*, it is suitable for the task of preserving the structure and integrity of the outliers in the clustering process. *Fifth*, the clustering is highly consistent with that perceived by human eyes. *Sixth*, the new approach provides a unified framework for scale-related clustering algorithms recently derived from many different fields, including estimation theory, information theory and statistical mechanics, radial basis function neural networks, and recurrent signal processing on self-organization feature maps.

To briefly describe the approach, we now relate the blurring process with the process of clustering. If $p(\boldsymbol{x})$ is a probability density function from which the data set is generated, then each blob is a connected region containing a relatively high density probability, separated from other blobs by a boundary with relatively low density probability. Therefore, each blob is a cluster. All blobs together produce a classification of a data space which provides a clustering for the data set with known distribution $p(\boldsymbol{x})$.

For a given data set $\boldsymbol{X} = \{x_i \in \Re^2 : i = 1, ..., N\}$, the empirical distribution for the data set $\boldsymbol{X}$ can be expressed as

$$\hat{P}_{emp}(\boldsymbol{x}) = \frac{1}{N}\sum_{i=0}^{N} \delta(\boldsymbol{x} - x_i) \tag{4.13}$$

The image corresponding to $\hat{P}_{emp}(\boldsymbol{x})$ consists of a set of light points situated at the data set, just like a scattergram of the data set. When we blur this image, we get a family of smooth images $P(\boldsymbol{x},\sigma)$ represented as follows

$$P(x, \sigma) = \frac{1}{N} \sum_{i=1}^{N} \frac{1}{\left(\sigma\sqrt{2\pi}\right)^2} \exp\left(-\left(\frac{\| x - x_i \|^2}{2\sigma^2}\right)\right) \tag{4.14}$$

The family $P(x, \sigma)$ can be considered as a Parzen estimation with Gaussian window function. At each given scale σ, the scale space image $P(x, \sigma)$ is a smooth distribution function, so the blobs and their centres can be determined by analyzing the limit of the solution $X(t, x_0)$ of the following differential equation

$$\begin{cases} \dfrac{dx}{dt} = \nabla_x P(x, \sigma) = \dfrac{1}{\sigma^2 N} \displaystyle\sum_{i=1}^{N} \dfrac{(x_i - x)}{\left(\sigma\sqrt{2\pi}\right)^2} \exp\left(-\left(\dfrac{\| x - x_i \|^2}{2\sigma^2}\right)\right) \\ x(0) = x_0. \end{cases} \tag{4.15}$$

When a distribution $p(x)$ is known, but contains noise or cannot be differentiated, we can also use the scale space filtering method to erase the spurious maxima generated by the noise. In this case, the scale-space image is

$$P(x, \sigma) = p(x)\ g(x, \sigma) = \int \frac{p(y)}{\left(\sigma\sqrt{2\pi}\right)^2} \exp\left(-\left(\frac{\| x - x_i \|^2}{2\sigma^2}\right)\right) dy \tag{4.16}$$

and the corresponding gradient dynamical system is given by

$$\begin{cases} \dfrac{dx}{dt} = \nabla_x P(x, \sigma) = \displaystyle\int \dfrac{p(y)(y - x)}{\left(\sigma\sqrt{2\pi}\right)^2 \sigma^2} \exp\left(-\left(\dfrac{\| x - x_i \|^2}{2\sigma^2}\right)\right) dy \\ x(0) = x_0 \end{cases} \tag{4.17}$$

When the noise in $P(x)$ is an independent white noise process, (4.16) provides an optimal estimate of the real distribution. By considering the data points falling into the same blob as a cluster, the blobs of $P(x, \sigma)$ at a given scale produce a pattern of clustering. In this way, each data point is deterministically assigned to a cluster via the differential gradient dynamical equation in (4.15) or (4.17), and thus our proposed scheme is a hard clustering method. As we change the scale, we get a hierarchical clustering.

Remark 4.2 Though neural networks can be used as stand-alone models for spatial data classification, it has been shown that their accuracy can be improved if they are used integratively with other methods, such as statistical techniques (Bruzzone, Prieto and Serpico 1999; Benediktsson and Kanellopoulos 1999), fuzzy sets (Tzeng and Chen 1998) and domain specific knowledge (Leung 1997a; Leung, Luo and Zhou 2000a; Murai and Omatu 1997). While neural networks have become a rather common tool for spatial classification and pattern analysis, the use of genetic algorithms for such a task is still in its infancy (see, for example, Tso and Mather 1999; Leung, Luo and Zhou 2000). The potential of evolutionary computation in this area still awaits exploration.

4.6 Rule Learning by a Radial Basis Function Neural Network

While the use of neural networks for classifying spatial data or images has become common practice, initial attempts are now being made to apply neural and other computational models to the learning of inferential rules, especially for intelligent spatial systems. This line of research, however, has yet to gain momentum.

Feedforward neural networks with the gradient-descent technique combined with backpropagation, among other rule-learning methods (see for example Kosko 1992) have been constructed to acquire fuzzy rules from examples (Lin and George 1991; Ishibuchi, Tanaka and Okada 1993; Nauck and Kruse 1993; Sulzberger, Tschichold-Gürman and Vestli 1993). Though these models appear to render good learning procedures, they may not converge at all, or have a very slow rate of convergence when the volume of learning examples is large, such as in remote sensing, and the variables involved are numerous.

To have a fast extraction of fuzzy and non-fuzzy IF-THEN rules, a RBF neural network has been developed for acquisition of knowledge from a large number of learning examples (Leung and Lin 1996). An RBF neural network is essentially a feedforward neural network with a single hidden layer. The transformation from the input space to the hidden-unit space is non-linear, while the transformation from the hidden-unit space to the output space is linear. The advantage is that the hidden units provide a set of radial basis functions constituting an arbitrary basis for the expansion of input patterns into the hidden-unit space. The RBF network differs from the multilayer perceptron in that it has one hidden layer and the transformation from the hidden layer to the output layer is linear.

Radial basis functions were first proposed for functional approximation (see the review by Powell 1992), but they have also been exploited for the design of neural networks (Broomhead and Lowe 1988) and fuzzy logic controllers (Steele et al. 1995). Here, we explain how a competitive RBF network can be constructed for the fast extraction of fuzzy and non-fuzzy rules for spatial inference (Leung and Lin 1966).

Without loss of generality, let the following be the ℓ-th rule of a set of IF-THEN rules to be learned:

$$\begin{aligned}&\text{If } x_1 \text{ is } \mu_{1\ell}, x_2 \text{ is } \mu_{2\ell}, \ldots, x_N \text{ is } \mu_{N\ell} \\ &\text{then } y_1 \text{ is } v_{\ell 1}, y_2 \text{ is } v_{\ell 2}, \ldots, y_P \text{ is } v_{\ell P}\end{aligned} \tag{4.18}$$

where $v_{\ell k}$, $k = 1, \ldots, P$, can be a real number, a fuzzy subset [e.g. a fuzzy number], or a binary number. We first assume that $v_{\ell k}$ is a fuzzy subset.

Let $\{(X^{(i)}, Y^{(i)}), i = 1, \ldots, M\}$, where $X^{(i)} = (x_1^{(i)}, \ldots, x_N^{(i)})$ and $Y^{(i)} = (y_1^{(i)}, \ldots, y_P^{(i)})$, be a subset of the M input-output patterns. Let S be the number of fuzzy subspaces partitioning each input and output space [for simplicity, we make the number of partitions in the input and output spaces the same; however, the method also holds for unequal number of partitions]. Let $[-\ell_j, \ell_j]$, $j = 1, 2, \ldots, N$, be the domain of input space j, and the centres of the fuzzy input subspaces be

$$C_j(1), C_j(2), \ldots, C_j(S) \qquad \text{for } j = 1, \ldots, N \tag{4.19}$$

Now, let $[-L_k, L_k]$, $k = 1, \ldots, P$, be the domain of output space k, and the centres of the fuzzy output subspaces be

$$v_k(1), v_k(2), \ldots, v_k(S) \qquad \text{for } k = 1, \ldots, P \tag{4.20}$$

The RBF network first determines the centres of the input and output subspaces through an unsupervised competitive learning procedure as follows:

- For each x_j input $x_j^{(1)}$, compute the degree of matching

$$d_{jr}^{(1)} = \left|x_j^{(1)} - C_j(r)\right| \qquad \text{for } r = 1, \ldots, S \tag{4.21}$$

- Select the best-match [winner] neuron and adjust its weight so that the node closest to $x_j^{(1)}$ has the greatest chance of winning [and thus achieving the clustering effect], that is let

$$\left|x_j^{(1)} - C_j(r_1)\right| = \min \; d_{jr}^{(1)} = \min\left|x_j^{(1)} - C_j(r)\right| \tag{4.22}$$

then

$$\Delta C_j(r) = \begin{cases} \eta_1\left[x_j^{(1)} - C_j(r)\right] & \text{if } r = r_1 \\ 0 & \text{otherwise} \end{cases} \tag{4.23}$$

where $\eta_1 \in (0,1)$ is a coefficient of learning. To prevent too large a fluctuation in the competitive clustering process, η_1 can be adjusted throughout. For example, we can take a larger value of η_1 at the beginning so that more nodes are involved in the competition. The value of η_1 can be subsequently reduced [e.g. we can take $\eta_1/\sqrt{k}$, where k denotes the number of sweeps of the training data set], so we only fine tune the relevant centre.

- Input $x_j^{(2)}$ and repeat the above steps until all M inputs are exhausted. The derived $C_j(1)$, …, $C_j(S)$ are then the winner centres of the fuzzy subspaces of x_j.

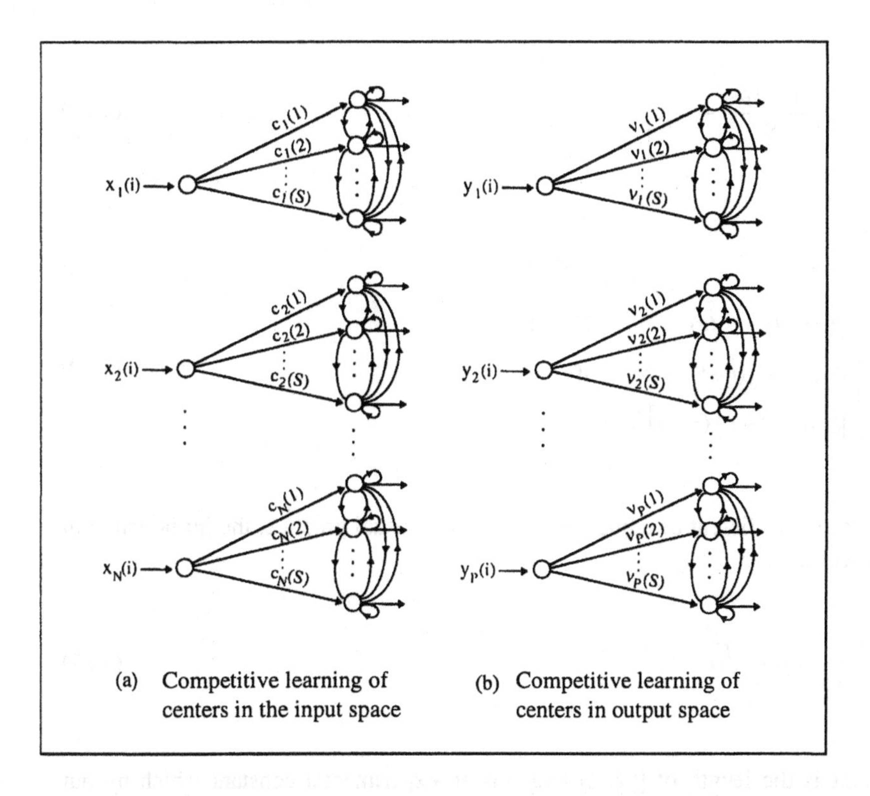

Fig. 4.5 Unsupervised competitive learning process (adopted from Leung and Lin 1996)

The above unsupervised competitive learning procedure can also be applied to the output spaces to derive the centres of the fuzzy output subspaces. The competitive process is depicted in Fig. 4.5(a) and Fig. 4.5(b). Here, the connection weights between the input and competitive layers are in fact the centres of the input and output subspaces. Within the RBF framework, the derived centres are actually the centres of the radial basis functions:

$$G(|x-c|) = \exp[-(x-c)^2/2\sigma^2] \tag{4.24}$$

which are actually the membership functions of the linguistic terms of the fuzzy IF-THEN rules in (4.18)

The spread of G is σ, which needs to be adaptively adjusted. For too large a σ, we would have too much overlapping of the fuzzy subspaces, resulting in unclear classification of rules. For too small a σ, we would have too condensed a radial basis function affecting the precision of computation. In place of the gradient descent procedure, which tends to overtune σ, we use the following procedure:

$$\Delta\sigma_{jr} = \alpha\left[\frac{h_{jr}\sqrt{\rho}}{S} - \sigma_{jr}\right] \tag{4.25}$$

where

$$h_{jr} = \begin{cases} C_j(r+1) - C_j(r) & \text{if } r = 1 \\ C_j(r) - C_j(r-1) & \text{if } r = s \\ \left[C_j(r+1) - C_j(r-1)\right]/2 & \text{if } 1 < r < S \end{cases} \tag{4.26}$$

and α is an coefficient of adjustment. Here, as is usual practice, the initial value of σ is selected as:

$$\sigma_r^0 = \frac{2\ell}{(S-1)\,S}\sqrt{\rho} \tag{4.27}$$

where 2ℓ is the length of $[-\ell, \ell]$ and ρ is an experimental constant which by our experience can be selected as follows:

$$\rho = \begin{cases} 3 & \text{if } S = 5 \\ 6 & \text{if } S = 7 \\ 21 & \text{if } S = 13 \end{cases} \tag{4.28}$$

For S=7, Fig. 4.6 depicts the classification of the subspaces of x_j after the unsupervised competitive learning and the adaptive adjustment of σ_{jr}. Since the centres of all subspaces can be determined as above, then the centres of the clusters in $\Re^N$ have coordinates

$$(C(i_1), \ ..., \ C_N(i_N)) \quad \text{for} \quad i_1, ..., i_N \in \{1, 2, ..., S\} \tag{4.29}$$

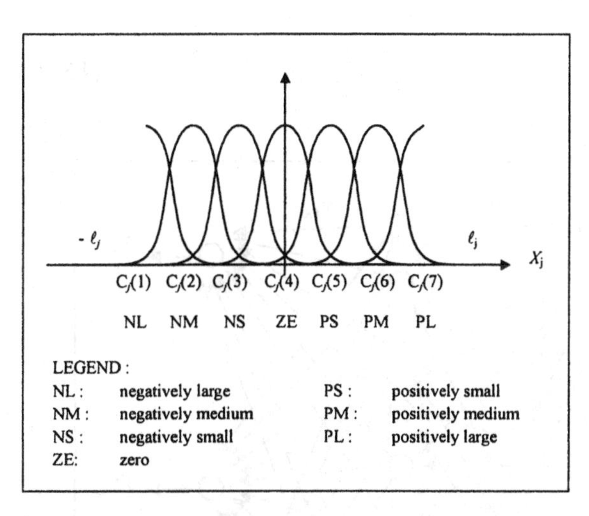

Fig. 4.6 Fuzzy partitions of input space X_j (Leung and Lin 1996)

The number of centres totals S^N and they are the centres of the radial basis functions of the nodes in the inference layer of the radial basis function network depicted in Fig. 4.7. Each of the S^N nodes corresponds to a fuzzy IF-THEN rule.

To recapitulate, the connection weights between the input and matching layers are real numbers specifying the centres of the fuzzy subspaces. The connection weights between the matching and inference layers are fuzzy numbers. The connection weights between the inference and output layers can be real numbers or fuzzy numbers, depending on the situation. Specifically, the connection weights between the ℓ-th node of the inference layer and all nodes in the matching layer are $\mu_{1\ell}, \ ..., \ \mu_{N\ell}$ with

$$\mu_{j\ell}(x) = \exp|-(x - C_j(r_{j\ell}))^2/2\,\sigma_j^2\, r_{j\ell}| \quad \text{for } j = 1, \ldots, N,\ r_{j\ell} \in \{1,\ldots, S\} \tag{4.30}$$

Let $C_l = (C_1(r_{1\ell}), C_2(r_{2\ell}), \ldots, C_M(r_{N\ell}))$ be the centre of the radial basis function corresponding to the ℓ-th node. Let $x_\ell^{(i)} = (x_1((i_\ell), \ldots, x_N(i_\ell))$ be the input, among M inputs, closest to C_ℓ, i.e.

$$\left\|x_\ell^{(i)} - C_\ell\right\| = \min_i \left\|x^{(i)} - C_\ell\right\| \tag{4.31}$$

Then, the connection weight between the ℓ-th node of the inference layer and the output layer can be taken as the output of the i_ℓ-th input, i.e.

$$v_{\ell 1} = y_1(i_\ell),\ v_{\ell 2} = y_2(i_\ell),\ \ldots,\ v_{\ell p} = y_{\ell P}(i_\ell) \tag{4.32}$$

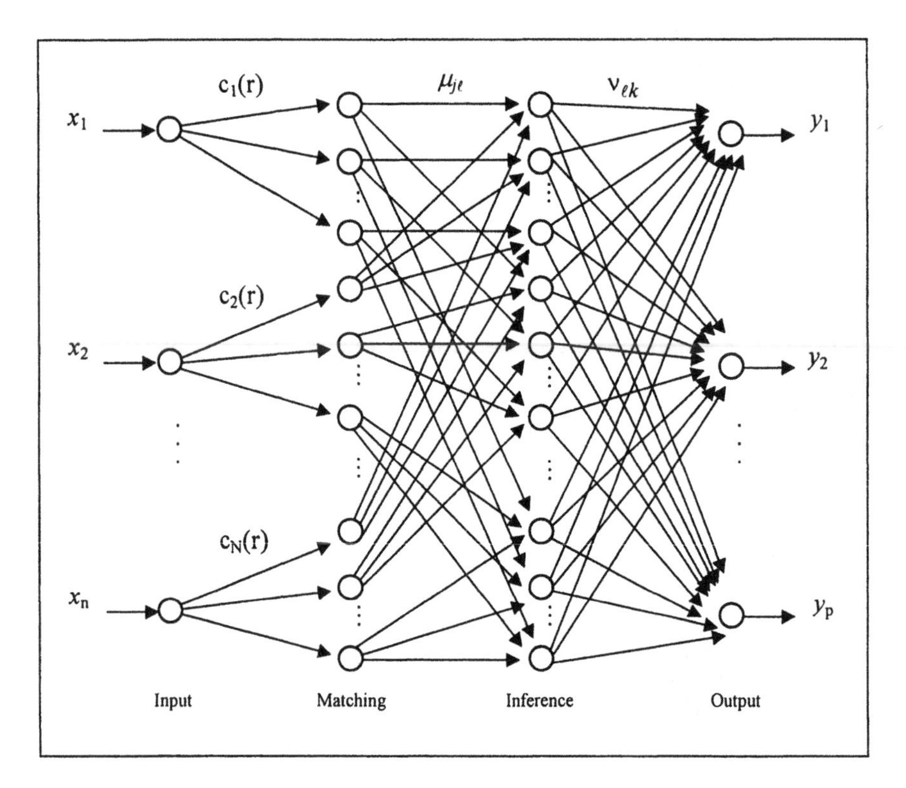

Fig. 4.7 Architecture of the RBF network for rule learning (Leung and Lin 1996)

Thus, the ℓ-th rule in (4.27) is extracted. Without further elaboration, $v_{\ell k}$ ($k = 1, \ldots, P$) can be a fuzzy number. If $v_{\ell k}$ is a real number, then the corresponding output is also a real number. By thresholding $v_{\ell k}$, we can also make a binary-valued output. Therefore, the RBF network can easily both extract fuzzy rules and non-fuzzy rules from a large number of examples. The performance of the network is supported by simulation results. Furthermore, it can perform on-line computation, rule deletion, and rule tidying which are essential features in rule-based systems (Leung and Lin 1996).

4.7 Rule Learning by a Hybrid Fuzzy Neural Network

Using neural networks for automatic learning from examples has been a common approach employed in the construction of computational systems. Backpropagation [BP] neural networks are models commonly used for learning and reasoning. However, neural information processing models generally assume that the patterns used for training a neural network are compatible. In some research (Monostori and Egresits 1994; Hernandez and Moore 1995), fuzzy neural networks are developed with stronger non-linear and imprecision mapping abilities compared with conventional BP networks. They appear to have promising application prospects in non-linear modelling, fuzzy identification and self-organizing fuzzy control for complex systems.

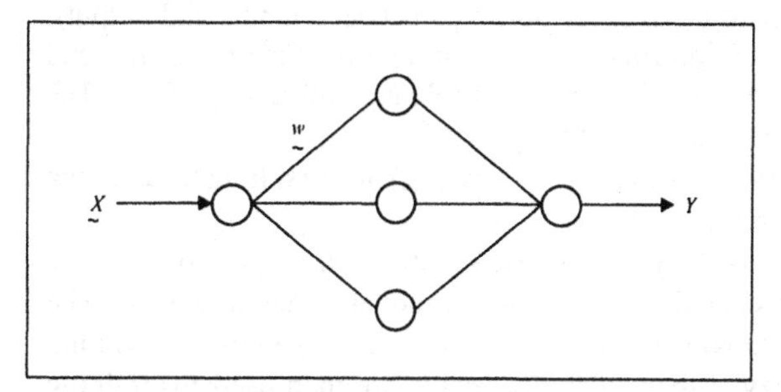

Fig. 4.8 BP network with fuzzy numbers (Huang and Leung 1999)

There are two common architectures for processing fuzzy information based on the conventional BP neural network. One of them is to use fuzzy numbers instead of real values as inputs and weights. This architecture (Lee and Lu 1994; Lee, Lu and Lin 1995) is called a fuzzy BP network and is shown in Fig. 4.8. Fuzzy BP networks perform non-linear mapping between fuzzy input vectors and crisp outputs. This type of BP network is not, however, able to process contradictory patterns, as it usually encounters the problem of convergence.

Another common fuzzy BP model is to generate fuzzy rules using the BP technique. In this architecture (Rhee and Krishnapuram 1993), as depicted in Fig. 4.9, an input node represents a fuzzy antecedent A_i and an output node means a fuzzy consequent B_i. Pattern (u,v) fires A_i and B_i in value $\mu_{A_i}(u)$ and $\mu_{B_j}(v)$. In other words, real values can be replaced by fuzzy memberships to train a conventional BP neural network. The fuzzy rules correspond to final weights which are larger than zero.

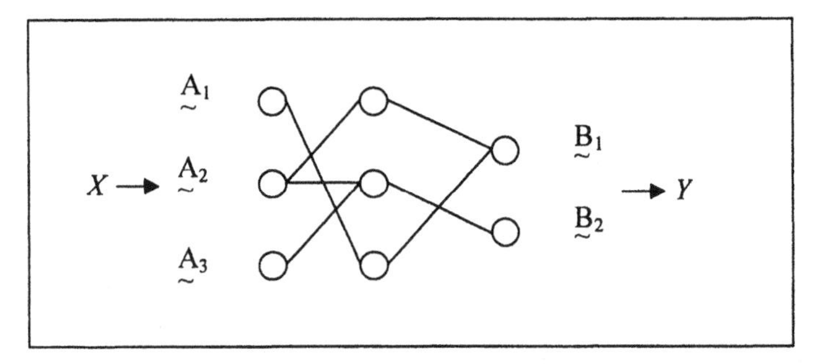

Fig. 4.9 Fuzzy-rule neural network model (Huang and Leung 1999)

Though the generation of fuzzy rules from training data by neural networks is more automatic, the method does not ensure that we can always find the rules with a given pool of experts or with a fixed set of data. To effectively process incomplete and contradictory data in order to unravel relationships between inputs and outputs, a hybrid model which integrates information-diffusion approximate reasoning and conventional BP neural network has been proposed (Huang and Leung 1999). The architecture of the hybrid model is depicted in Fig. 4.10.

In the hybrid model, observations (m_1, g_1), ..., (m_n, g_n) are first transformed, via the information-diffusion technique, into new patterns $(m_1, \hat{g}_1)$, ..., $(m_n, \hat{g}_n)$. A conventional backpropagation [BP] neural network is then employed to learn the relationship between the isoseismal area, g, and earthquake magnitude, m. The information-diffusion approximate reasoning serves as a pattern smoothing mechanism for the set of fuzzy if-then rules. Its results are then used for training the BP neural network.

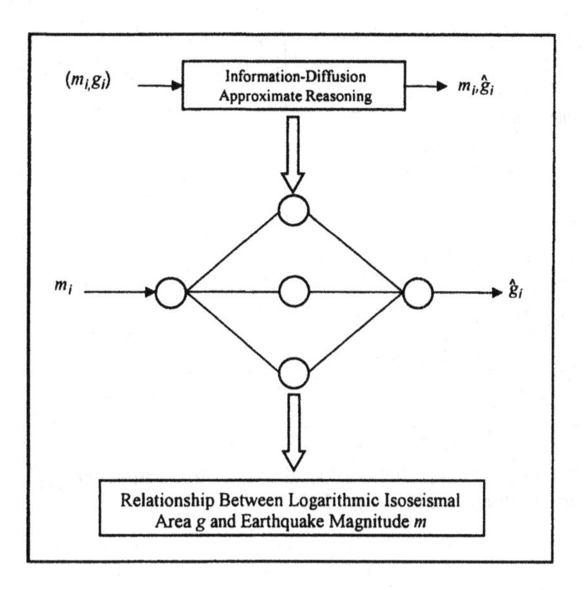

Fig. 4.10 Architecture of the hybrid model integrating information-diffusion approximate reasoning and conventional BP neural network (Huang and Leung 1999)

Suppose we have n observations (m_1, g_1), ..., (m_n, g_n). Using the information diffusion technique, we can get n fuzzy if-then rules $A_1 \rightarrow B_1$, ..., $A_n \rightarrow B_n$ as shown in Fig. 4.11. Now, if a crisp input value m_0 [the antecedent] is known, then we need a way to infer the consequent g_0 from m_0 and R_i. In practice, U is generally discrete, so m_0 is not just equal to some value in U. We can employ the information distribution formula (4.33) to get a fuzzy subset as an input

$$\mu_{m_0}(u_j) = \begin{cases} 1 - |m_0 - u_j| / \Delta & \text{if } |m_0 - u_j| \leq \Delta \\ 0 & \text{if } |m_0 - u_j| > \Delta \end{cases} \tag{4.33}$$

where $\Delta = u_{j+1} - u_j$. A fuzzy consequent $\underset{\sim}{g_0}$ from $\underset{\sim}{m_0}$ and R_i is then obtained as

$$\mu_{g_0}(v) = \sum_u \mu_{m_0}(u)\ \mu_{R_i}(u, v) \tag{4.34}$$

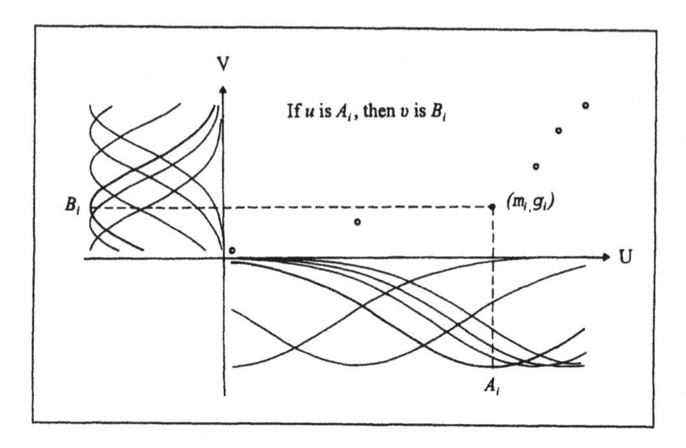

Fig. 4.11 Production of the fuzzy rule 'If u is A_i, then v is B_i' through observation (m_i, g_i) by the information diffusion technique (Huang and Leung 1999)

When we defuzzify $\underset{\sim}{g}_0$ into a crisp output value, there is no operator which can avoid system error. In order to get directly crisp output value, we only need to change the magnitude component into fuzzy subsets.

$$\mu_{B_i}(v) = \begin{cases} 1 & \text{if } v = g_i \\ 0 & \text{if } v \neq g_i \end{cases} \tag{4.35}$$

and

$$\mu_{R_i}(u, v) = \begin{cases} \mu_{A_i}(u) & \text{if } v = g_i \\ 0 & \text{if } v \neq g_i \end{cases} \tag{4.36}$$

Therefore,

$$\mu_{g_0}(v) = \begin{cases} \sum_u \mu_{m_0}(u)\ \mu_{A_i}(u) & \text{if } v = g_i \\ 0 & \text{if } v \neq g_i \end{cases} \tag{4.37}$$

Let

$$w_i = \sum_u \mu_{m_0}(u)\,\mu_{R_i}(u) \tag{4.38}$$

In fact, w_i is the possibility [weight] that consequent g_0 may be g_i. Then, to integrate all results coming from $R_1, \ldots, R_n$, the relevant output value g_0 becomes

$$g_0 = \frac{\sum_{i=1}^{n} w_i\, g_i}{\sum_{i=1}^{n} w_i} \tag{4.39}$$

The procedure comprising Equations (4.33)-(4.39) is called 'information-diffusion approximate reasoning' whose system architecture is depicted in Fig. 4.12. Any observation (m_i, g_i) of the sample can thus be changed into a new pattern $(m_i, \hat{g}_i)$ via information-diffusion approximate reasoning. It is apparent that the new patterns must be smoother.

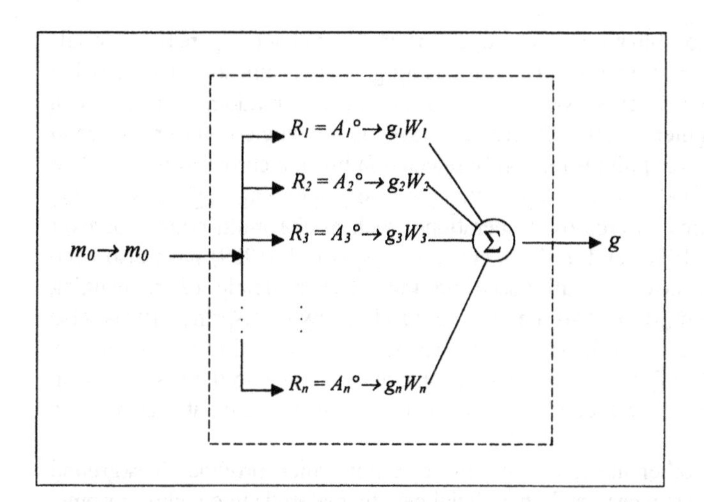

Fig. 4.12 System architecture of information-diffusion approximate reasoning. '°' is the compositional rule defined by Equation (4.36) (Huang and Leung 1999)

4.8 Rule Acquisition by Genetic Algorithms – The SCION System

Genetic algorithms are domain independent global search methods which are modelled after the mechanics of natural evolution within populations and species via reproduction, cross breeding, mutation, competition, and selection. A genetic algorithm generally starts with an initial population, i.e. a set of individuals. Each individual is a candidate solution [e.g. a rule, a logical operation, or a neural network topology] consisting of a single chromosome with genes as elements and alleles as values in varying positions. A chromosome is usually represented as a binary string with specific length [reflecting the desirable degree of precision for a problem]. A Darwinian notion of fitness [goal-oriented evaluation] is then employed to determine selective reproduction. Chromosomes are selected, usually by a probabilistic procedure [e.g. the roulette wheel selection] with respect to their fitness values [evaluated by a fitness or evaluation function] for reproduction. Those with higher relative fitness values will have a higher chance of being selected to produce future generations [successive populations]. Genetic operators such as crossover and mutation are employed to generate the next population. They are basically operators which determine the genetic make-up of offspring from the genetic materials of the parents.

Crossover is a recombination genetic operator which exchanges genetic materials from two parent chromosomes to produce offspring. For chromosomes with string length n there are $n-1$ crossover points to be chosen randomly to break a chromosome into segments. The simplest crossover is the single-point crossover in which only one crossover point is randomly selected to break a chromosome into two segments. Offspring are produced by exchanging corresponding segments of two parents. More complicated crossover operations, such as the multi-point crossover (De Jong and Spears 1992) and uniform crossover (Syswerda 1989) have also been proposed. They are, however, all based on the same principle of exchanging corresponding segment(s) of two parents to produce two offspring. [It is also permissible to produce one offspring from two parents.] For uniform crossover, a template is used to specify which parent will contribute its value to the first offspring at a particular position. The second offspring will receive the value at that position from the other parent.

Mutation, on the other hand, is a genetic operator which provides background variation and introduces occasionally beneficial genetic materials into a chromosome. The task is essentially done by altering the value(s) of a randomly selected position(s) in a string. Offspring thus produced then constitute the next generation and are subjected to the validity and fitness test. The iterative process continues until a desirable population is obtained.

The basic structure of a genetic algorithm can be summarized as follows:

```
begin
  t = 0
  initialize population P(t)
  evaluate P(t)
  while (not termination-condition) do
  begin
    t = t+1
    select P(t) from P(t-1) based on the fitness of the
    individuals in P(t-1)
    generate (by crossover and mutation) structures in P(t)
    evaluate P(t)
  end
end
```

Though selective reproduction, crossover and mutation are simple mechanisms employed in genetic algorithms, they appear to be powerful and robust search techniques for knowledge acquisition. Mathematically, a string of length n is a point [intersection of hyperplanes], i.e. a vector, in the search space. Without knowledge of the search space [such as the gradient information and parameters], genetic algorithms just need to know the quality of the solution produced by each parameter in order to seek the solution by considering a population of points, instead of a point-by-point search. They are thus good for knowledge acquisition in large and complex search spaces. They can make use of accumulated information about an initially poorly understood search space to bias subsequent search into advantageous subspaces. To enhance efficiency, domain knowledge, if available, can be employed to choose the initial population, internal representation, genetic operators, feedback mechanisms [e.g. reward-and-punishment scheme], and space to be searched.

There are many ways production rules can be represented by chromosomes in genetic algorithms. The rules represented by a population of chromosomes are then evolved by a genetic algorithm to cover all learning examples to solve the given problem. In general, we can either treat rule sets as entities or rules as entities. Treating rule sets as entities, an entire rule set is represented as a string of variable length (Smith 1983; Schaffer 1985). Within this framework, individual rules are represented as genes and the entire program as strings of these genes. The genetic algorithm will maintain a population of candidate rule sets and employ selective reproduction and genetic operators to produce new generations of rule sets. Crossover, for example, can be used to produce new combinations of rules and mutation can be used to produce new rules. However, evolving rule sets of varying length and complexity by canonical genetic algorithm is not too effective. Several extensions have been proposed to alleviate the problem. The classifier system is a typical example.

Treating rules as entities, members of the classifier systems are individual rules [classifiers] and a rule set is represented by the entire population (Holland and Reitman 1978; Booker, Goldberg and Holland 1989). Classifier systems comprise a set of rules which manipulate an internal message list. A fixed-length approach is commonly employed to represent the left-hand-side [LHS] of each rule consisting of a pattern which matches message to be posted on the message list. The right-hand-side [RHS] of each rule specifies a message to be posted on the message list if the rule is to be fired. Perturbation of a classifier system occurs with the arrival of one or more detector messages indicating a change in the environment. A sequence of rule firing is triggered by the content change of the message list. A reward-punishment system [a feedback mechanism] is used to indicate favourable and unfavourable states. A bucket brigade (Holland 1986) is used to distribute payoffs to active rules in sequences resulting in rewards. In brief, the classifier system, a specific type of adaptive learning system, is a massively parallel, message-passing, rule-based system which learn through credit assignment by the bucket brigade algorithm and rule discovery by the genetic algorithm. It is, however, not quite suitable for extracting explicit rules for a knowledge base.

Being a global multipoint search method, genetic algorithms can be employed to effectively and automatically learn rules from examples. Acquired rules can be translated into the symbolic form suitable for the conventional rule-based expert system shell to make inference. To learn rules for the rule-based expert system shell, a platform code-named *SCION* has been built for the development of genetic-algorithm-based applications (Leung et al. 1992a, b). Two novel ideas, namely, token competition and rule migration were introduced in conjunction with crossover and mutation.

The basic idea of *SCION* is to use a genetic algorithm to learn rules from examples given as a set of data. An initial population of rules is germinated randomly and then subjected to *rule evaluation*, a competitive process in which the rules are ranked according to their fitness obtained from comparing with the set of training examples. The weaker rules are eliminated. The remaining elite rules are used by the genetic algorithm to produce offspring [new rules] by *crossover* and/or *mutation.* To complete an iteration [an evolution cycle], the new-born rules join the competition [rule evaluation] after being treated by a *rule tidying* process which prunes redundant components in each new rule. The cycle stops when the population of rules satisfies certain criteria, or when a preset number of iterations is reached.

The crossover operation involves two rules. The randomly selected part from each rule are joined by a random Boolean relation to give new offspring. For example, '(X2 < > X3) AND (NOT (X1 < 4))' can crossover with '(X1 + X2 = 9) OR (X4 < 5)' to give '(NOT (X1 < 4)) OR (X1 + X2 = 9)', where OR is randomly generated. The mutation operator randomly selects a candidate from the elite rules and performs a grow [add], slim [cut] or a change operation on one of its randomly selected opcodes or operands.

The overall system flows of *SCION* is depicted in Fig. 4.13 and is explained here briefly.

(a) *Rule Representation.* In order to increase efficiency, we employ a new structure, a chain of duples, instead of the tree representation. A duple is an entity containing two values which are associated with an attribute of a rule. The values are the lower and upper bounds of the attribute. Suppose a rule containing three duples has the following form:

((4, 7), (3, 9), (11, 20))

The equivalent Boolean form is:

IF $(4 \le X_1 \le 7)$ AND $(3 \le X_2 \le 9)$ AND $(11 \le X_3 \le 20)$ THEN CLASS = 1. (4.40)

The advantage of using duples is that a greater resemblance between the rule structure and its biological counterpart can be established. The chain of duples is an analogy of a chromosome and a duple is an analogy of a gene. Also, with duple representation, genetic operators like crossover and mutation can be made faster and simpler because the simple array data structure storing the duples resembles the gene-chromosome relationship.
A crossover operator involves only the splitting of two parents and then the recombination of their two separated segments. The mutation operator involves only the random changing of an arbitrary duple. Thus the learning process can be speeded up. The simple representation only allows AND and $\le$ relationships in a rule, but almost all relationships can be simulated by a set of simple rules. For example, an OR relationship can be modelled by two separate rules in the inference process.

(b) *Rule Evaluation and Sorting.* This module is used to calculate the strength of all rules by keeping track of their hit scores and false alarms as follows:

for each class i
read sample data until end of file
for each rule in class i
if it can classify the data
increase its hit score by 1
for each rule in other class
if it can classify the data
increase its false alarm by 1.

All rules within one class are sorted in descending order by their strengths. A class can consist of two data zones with contrasting sizes. There is a risk that the smaller zone will be ignored. This is because any rule landing on the larger zone will enjoy a good score in the hit-false strategy, even though it may have a false alarm. Those rules landing on the smaller zone will suffer from limited credit, even though they have no false alarms at all. These unfortunate rules will then face the risk of being expelled, since competition for survival is keen. This small zone will, as a result, be left out, and consequently perfect performance can never be reached. However, under the hit/false instead of the conventional hit-false scoring strategy, rules in small clusters still have the chance to get high credit if they have proportionally less false alarms than those rules in bigger clusters. Therefore, the small clusters are still accounted for. To avoid the dividing-by-zero error when 'false' = 0, a strength adjustment is added to the denominator, making the formula to be hit/[false + adjust].

(c) *Token Competition.* The inbreed mating poses a great threat to the robustness of a genetic algorithm. As this phenomenon propagates, the whole population will gradually degenerate into a set of homogeneous rules. Then the evolution will fall into a vicious circle as the homogeneous parent rules reproduce homogeneous children rules. Token competition is designed to remove this hindrance. In this mechanism, each sample datum is regarded as a token. Once a rule can correctly classify this datum, it will seize the token so that rules to follow cannot get it. After the whole sample database is parsed, those rules with no token in hand will be killed. The priority of receiving tokens is determined by the strength of the rules. Strong rules are encouraged to seize as many tokens as they can, leaving the weak rules starving. As a result, the whole sample data set can be represented by only a small set of strong rules, making the final production rules more concise. In addition, more room can be saved for reproducing more children rules by cutting the redundant parent rules without affecting the performance. The robustness of the genetic algorithm is enhanced as more search points are explored by this larger set of children rules. Thus, the token competition module determines the number of tokens each rule can get. The rules are assumed to be sorted already. Token allocation is determined as follows:

for each class i
read sample data until end of file
give token to the first rule in class i classifying it.

(d) *Redundant and Weak Rule Elimination.* After token competition, each rule will get its own tokens. Those which cannot get any token are classified as redundant and will be eliminated and imprisoned. If the size of the rule set still exceeds the allowed quota of parent rules after redundant rule elimination, the excess rules [the weaker ones] will be eliminated.

(e) *Rule Migration.* Rule Migration is a novel idea in genetic algorithms. The concept behind this idea is simple. During the evolution process, a weak species in one class may be a strong species in another class, and can be preserved rather than simply discarded. This idea comes from the observation that the classes themselves are disjointed. Therefore a rule can only score high in its should-be class. Thus, the good rule in a certain class needs not migrate to other classes, for it must have low scores in other classes. However, a poor rule in one class may score well in other classes. Without migration, this valuable rule for other classes may be discarded and wasted.
Immigrant rules can be treated as offspring in the reproduction process. If after migration, there is still room for crossover and mutation, original rules in that class will reproduce to fill up the population size. If an immigrant rule behaves better than the newly produced offspring, then its last immigration is said to be successful. Otherwise, it has less contribution to this class than the offspring from the original population, so after one generation it will be discarded. Since rule migration happens earlier than rule reproduction, the more immigrant rules migrate to a class, the less offspring the original rules in

that class can reproduce. So, an immigration quota is required to set the maximum number of immigrants to a class.

(f) *Reproduction by Genetic Operators.* After two severe screening procedures, all the survived rules will become potential parents. Crossover is performed first. Two parents are involved in contributing their 'genes' to form the children rules. The selection of parents is a process by which a rule with greater strength has a greater chance of being selected. The quota of reproducing children rules by the crossover operator is determined by the crossover ratio supplied by users. The rest of the space belongs to the mutation process.
In crossover, the selected parent rules will duplicate themselves first. Then each copy will be split into two sections with the cut-off point randomly selected. The section before the cut-off point is detached from the rest of the body. These sections will then be exchanged and recombination takes place, giving two new children rules. Suppose the cut-off point is at the 4th position for the following two copies of the parent rules:

(1, 10), (4, 11), (20, 30)
(58, 90), (7, 40), (1, 5)

Then, two children rules are obtained as follows:

((1, 10), (4,40), (1, 5))
((58, 90), (7, 11), (20, 30)).

Each recombination must be checked to prevent inappropriate matching between a large lower bound and a small upper bound. Also, the children rule produced must not resemble any existing rules. If any duplication is found, the crossover operator is reapplied again until a successful mating is achieved or until too much failure is encountered. The latter case may suggest that the combinations of genes of the parents are exhausted. Then the excess quota will be granted to the mutation operation. The mutation operator will just select randomly a value to be changed. The partner of the value in the duple will be referenced to guarantee that no lower bound is greater than the upper bound. Suppose mutation takes place at position 2 of the following rule:

((4, 23), (17, 34), (1, 9).

The child rule is then obtained as:

((4, 23), (17, 91), (1, 9)).

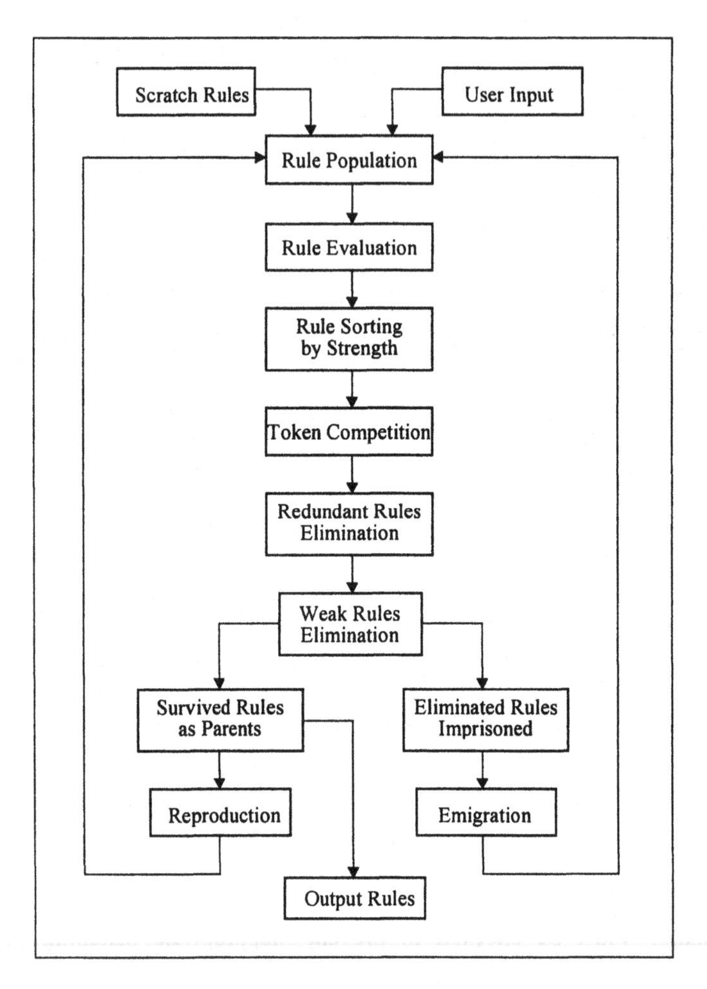

Fig. 4.13 The overall system flows of SCION (from Leung 1997)

4.9 Fuzzy Rule Acquisition by Genetic Algorithms – The GANGO System

To acquire fuzzy rules for spatial inference or pattern analysis, we can consider the problem as classifying a pattern vector $\boldsymbol{a} = (a_1, ..., a_d)$ from a d-dimentional pattern space $A \in \mathrm{R}^d$ into M classes. The task is to design a computational device that can output a class index $i \in \{1, \ldots, M\}$ for each input vector in the pattern space. The method of fuzzy logic is instrumental in pattern classification under imprecision. The construction of a fuzzy system for classification problems involves three basic aspects: (i) determination of fuzzy inference method, (ii) fuzzy partition of the pattern space into fuzzy subspaces, and (iii) generation of a set of fuzzy rules. We can adopt the method of simple fuzzy grid to partition the pattern space A (Ishibuchi et al. 1995). An example of a fuzzy grid partition is shown in Fig. 4.14 where the two-dimensional pattern space is divided into 9 fuzzy subspaces A_{ij}, $1 \leq i, j \leq 3$. Other more complicated partitioning methods are possible. Among all fuzzy inference methods, in our fuzzy system we only use a variation of the so-called *new fuzzy reasoning method* (Cao, Kandel and Li 1990; Park, Kandel and Langholz 1994) as a basis of study.

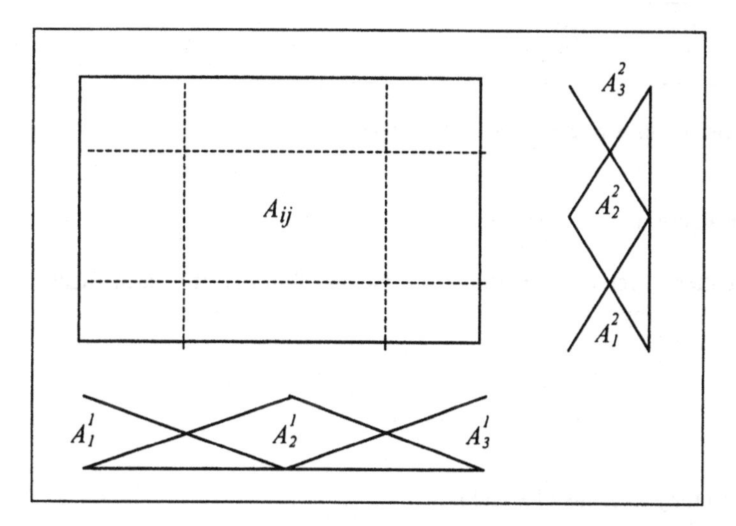

Fig. 4.14 Fuzzy grid-partitioning of a pattern space (from Leung et al. 1997b)

We depict our fuzzy system schematically in Fig. 4.15. The vector $\boldsymbol{X} = (x_1, ..., x_N)$ has component x_i, $i = 1, \ldots, N$, being the degree of membership of the input pattern belonging to the ith fuzzy partition space A_i, i.e. $x_i = A_i(\boldsymbol{a})$, N is the number of fuzzy subspaces of the fuzzy partition; $\boldsymbol{Y} = (y_1, ..., y_M)$ is a vector with y_i denoting the degree of membership of the ith class; and $\boldsymbol{w} = (w_{ij})$ is an N by M fuzzy relationship matrix. The output b is an integer from $\{1, \ldots, M\}$ indicating the class number of the input pattern.

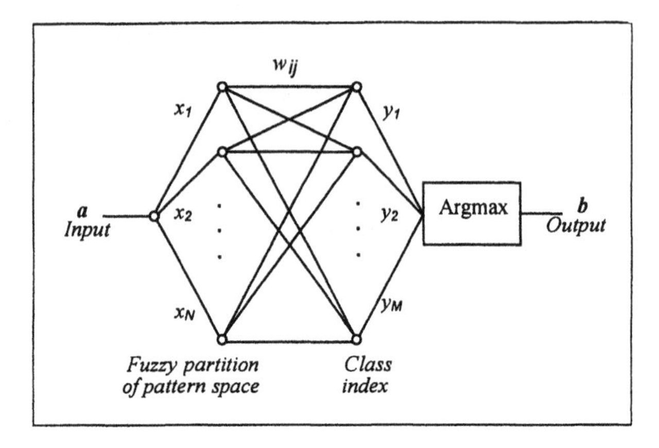

Fig. 4.15 A schema of fuzzy rule set (Leung et al. 1997b)

The inference algorithm is given as follows:

- For an input pattern vector ***a***, determine the membership of ***a*** for each fuzzy partition A_i, $1 \le i \le N$, by

$$x_i = A_i(\boldsymbol{a}).$$

- Calculate the vector ***y*** in terms of ***x*** and ***w***:

$$y_i = \sum_{i=1}^{N} x_i w_{ij}, \quad 1 \le j \le M.$$

- Find an index i_m such that

$$y_{i_m} = \max_{1 \le i \le M} y_i$$

and let the output b equal i_m.

This fuzzy system actually consists of N by M fuzzy rules, each of which can be identified in the form of 'IF ***a*** is A_i, then ***a*** belongs to class j with certainty w_{ij}'.

A. The Training Task as an Optimization Problem

We follow the method of Ishibuchi et al. (1995) to formulate the training task. Let $(\boldsymbol{a}_p, i_p)$, $p = 1, \ldots, L$, be L training patterns where $\boldsymbol{a}_p$ is the pattern vector and i_p is the class index of $\boldsymbol{a}_p$. The training task of the fuzzy system described above can be formulated as the following optimization problem:

Problem: Find a fuzzy relationship matrix $\boldsymbol{w}$ to maximize the function

$$f(\boldsymbol{w}) = \sum_{p=1}^{L} I_{i_p}(b_w(\boldsymbol{a}_p)) \tag{4.41}$$

where

$$I_i(b) = \begin{cases} 1 & \text{if } b = i \\ 0 & \text{otherwise} \end{cases}$$

and $b_w(\boldsymbol{a}_p)$ are the output of the fuzzy system with relationship matrix $\boldsymbol{w}$ when the input is $\boldsymbol{a}_p$. Since $f(\boldsymbol{w})$ is a function with continuous variables and discrete values, neither numerical optimization methods nor non-numerical methods, such as simulated annealing, can be successfully used to solve the above problem. So it is natural to resort to the more general and powerful method of genetic algorithms.

The training task of the fuzzy classification system is to find a fuzzy relationship matrix $\boldsymbol{w}$ that solves the optimization problem in (4.41). In order to use the genetic algorithm to solve the problem, we must first determine an encoding scheme that can transfer the fuzzy relationship matrix $\boldsymbol{w}$ into a binary string. A conventional [and also clumsy] encoding method is to represent each element in $\boldsymbol{w}$ in its binary form and then to combine these binary strings into a large string. However, because of the nature of our new genetic algorithm and the training problem of the fuzzy system, we can here adopt a new strategy. Although the w_{ij}'s are deterministic values, it is advantageous to consider them to be the expectation of some random variables. This viewpoint has been proven useful in the study of complex network systems, e.g. the recent trends in the study of [deterministic or stochastic] neural networks using probability models (Amari 1990, 1995). To begin with, we introduce the following concept.

Definition 1. A fuzzy system $\boldsymbol{w}$ is called a fuzzy system with crisp fuzzy relationship matrix if all the w_{ij}'s take on the value 0 or 1.

Definition 2. A fuzzy system $\boldsymbol{v}$ is called a fuzzy system with random and crisp fuzzy relationship matrix if each element w_{ij} of $\boldsymbol{v}$ is a 0-1 random variable.

Given a fuzzy system V with random and crisp fuzzy relationship matrix, let $w_{ij} = E(v_{ij}) = P\{v_{ij} = 1\}$. The fuzzy system $\boldsymbol{w} = (w_{ij})$ will be called the mean fuzzy system of $\boldsymbol{v}$. Conversely, any fuzzy system $\boldsymbol{w}$ can be viewed as the mean fuzzy system of some fuzzy system with random and crisp fuzzy relationship matrix. For convenience, we view the N by M matrices $\boldsymbol{w}$ and $\boldsymbol{v}$ as vectors whose components are still indexed by subscript i, j. For example, we treat $\boldsymbol{w} = (w_{ij}, 1 \le i \le N, 1 \le j \le M)$ as a vector of $N \times M$ dimension with the $((i-1)M + j)$th components being w_{ij}. In this way, any [random] and crisp relationship matrix may be viewed as a [random] binary string. Now return to the problem of encoding a fuzzy system $\boldsymbol{w}$ for our new genetic algorithm. We view the fuzzy system $\boldsymbol{w}$ involved in the training task in problem (4.41) as the mean fuzzy system of a fuzzy system $\boldsymbol{v}$ with random and crisp fuzzy relationship matrix. To find w_{ij} is equivalent to finding the parameter of the 0-1 distribution of v_{ij}, $P\{v_{ij} = 1\}$. In our algorithm for the training of the fuzzy system, we use the crisp relationship matrices $\boldsymbol{v}$'s as the individuals while the corresponding fuzzy re ionship matrices $\boldsymbol{w}$'s are given by the expectations of the random and crisp :lationship matrices corresponding to the individuals in the algorithm, that is indeec ne parameters of the 0-1 distribution of v_{ij}'s, $P\{v_{ij} = 1\}$.

Having specified the encoding scheme, we now summarize our algorithm for the training of fuzzy classification systems as follows (see Leung, Gao and Zhang 1997b):

Step 1: Randomly generate T fuzzy systems with crisp relationship matrix $\{\boldsymbol{v}(t)\}_1^T$. Compute the fitness of $f(\boldsymbol{v}(t))$ according to (4.41). Compute the characteristic $(F(0), F_{ij}^1(0))$ of the population $\{\boldsymbol{v}(t)\}_1^T$ according to:

$$F(\boldsymbol{X}(t-1)) = \sum_{i=1}^{N} f(X_i(t-1)) \tag{4.42}$$

$$F_j^1(\boldsymbol{X}(t-1)) = \sum_{i \in I_1(j)} f(X_i(t-1));\ I_1(j) = \{1 \le i \le N;\ x_{ij}(T-1) = 1\}. \tag{4.43}$$

For each pair (i, j) compute

$$p_{ij}^{(0)}(1) = \frac{F_{ij}^1(0)}{F(0)} + \left(1 - 2\frac{F_{ij}^1(0)}{F(0)}\right) P_m$$

and let $w_{ij}(0) = p_{ij}^{(0)}(1)$. Set $t = T$ and $k = T$.

Step 2: Sample the zero-one distribution $p_{ij}^{(t)}(.)$, $1 \le i \le M$, $1 \le j \le N$, with parameters $p_{ij}^{(t)}(1)$, to get an individual $v(t+1) = (v_{ij}(t+1))$.

Step 3: If $f(v(t+1)) < (F(0)/k)$, set $t = t + 1$ and return to *Step 2*; else update the characteristic $(F(t), F_{ij}^{1}(t))$, according to:

$$F(X(t)) = F(X(t-1)) + f(X) \tag{4.45}$$

$$F_j^1(X(t)) = \begin{cases} F_j^1(X(t-1)) + f(X) & \text{if } x_j = 1 \\ F_j^1(X(t-1)) & \text{if } x_j = 0 \end{cases} \tag{4.46}$$

to get the new characteristic $(F(t+1), F_{ij}^1(t+1))$. Set $k = k + 1$.

Step 4: For each pair (i, j) compute

$$p_{ij}^{(t+1)}(1) = \frac{F_{ij}^1(t+1)}{F(t+1)} + \left(1 - 2\frac{F_{ij}^1(t+1)}{F(t+1)}\right)P_m \tag{4.47}$$

and let $w_{ij}(t+1) = p_{ij}^{(t+1)}(1)$. Set $t = t + 1$.

Step 5: Repeat *Step 2 – Step 4* until some stopping criterion is met.

B. The Reduction of the Number of Fuzzy Rules in the Trained Fuzzy System

In real-world problems, the number of possible rules may be huge due to the high dimension of the pattern space. To improve computational efficiency and to obtain a sensible system, we need methods to eliminate some irrelevant rules so that we can have compact fuzzy system. The irrelevant rules essentially fall into two categories: the dummy rules and the inactive rules. Recall that the fuzzy rules in our fuzzy systems take the form:

If $\boldsymbol{a} \in A_i$, then $\boldsymbol{a}$ belong to class j with certainty w_{ij}.

Let $(\boldsymbol{a}_p, i_p)$, $p = 1, \ldots, L$, be the training patterns. A fuzzy rule is called a α-level dummy rule if $\Sigma_{i_p=j} A_i(\boldsymbol{a}_p) < a$. A fuzzy rule is called a β-level inactive rule if $w_{ij} < \beta$. Both the dummy rules and the inactive rules have little or no effect on the performance of the fuzzy systems, and should be eliminated.

(i) *Fitness Reassignment Strategy for the Elimination of Dummy Rules*

Our point of departure is the consideration that although a fuzzy system has an overall fitness, different fuzzy rules in the fuzzy system have different contributions to that overall fitness. For example, the dummy rules make little or no contribution to the performance [fitness] of a fuzzy system. Our strategy for the elimination of the dummy rules is to discourage dummy rules in the course of evolution by reassigning the fitness to the dummy rules. This is possible only in our new genetic algorithm framework, since it operates on the level of the components [i.e. genes or fuzzy rules], while selection in conventional genetic algorithms is done on the individuals [i.e. fuzzy systems] level. To implement the reassignment of fitness to the dummy rules in the training algorithm, all that needs to be changed is the updating scheme of $F_{ij}^1(t+1)$ in *Step 3*. For each $1 \leq i \leq M$, $1 \leq j \leq N$, define the weight of reassignment as

$$r_{ij} = \begin{cases} 0 & \text{if } \sum_{i_p = j} A_i(\boldsymbol{a}_p) < \alpha \\ 1 & \text{else} \end{cases} \tag{4.48}$$

where α is a small scalar. The updating scheme of $F_{ij}^1(t+1)$ becomes

$$F_{ij}^1(t+1) = \begin{cases} F_{ij}^1(t) + r_{ij}\, f(v(t+1)) & \text{if } V_{ij} = 1 \\ F_{ij}^1(t) & \text{if } V_{ij} = 0 \end{cases} \tag{4.49}$$

(ii) *Weight Truncation Strategy for the Reduction of Inactive Rules*

Let $\boldsymbol{w} = (w_{ij})$ be the fuzzy relationship matrix of a trained fuzzy system. As has been explained previously, this fuzzy system consists of $N \times M$ fuzzy rules of the form

If $\boldsymbol{a}$ is A_i, $\boldsymbol{a}$ belongs to class j with certainty w_{ij}.

Moreover, based on our probabilistic interpretation of w_{ij} in the encoding scheme, we can view w_{ij} as the conditional probability that a pattern belongs to class j, given that the pattern is in A_i. Otherwise w_{ij} can also be viewed as the probability that the rule 'If A_i then j' is active in the fuzzy system. We can reduce the number of fuzzy rules by eliminating those rules whose active probability w_{ij} is small. This is done by setting to zero the w_{ij}'s that are smaller than a small scalar [threshold]. Formally, let $\boldsymbol{w} = (w_{ij})$ be the fuzzy relationship matrix of a trained fuzzy system and let $0 < \alpha < 1$ be the threshold. Define a new fuzzy relationship matrix $\boldsymbol{w}^{\alpha} = (w_{ij}^{\alpha})$ by

$$w_{ij}^{a} = \begin{cases} w_{ij} & \text{if } w_{ij} > a \\ 0 & \text{if } w_{ij} \le a \end{cases} \tag{4.50}$$

The number of active fuzzy rules in the fuzzy system with fuzzy relationship matrix w^{α} is thus less than that in the original fuzzy system. This gives a tight set of rules with a sensible interpretation.

The GANGO system has been employed to extract rules and form a classification system for remotely sensed data. The automatic knowledge acquisition process was completed in relatively short training time (Leung, Gao and Zhang 1997)

4.10 Conclusions

Some of the major developments and applications of neural and evolutionary computations in spatial classification and knowledge acquisition have been examined in this chapter. It has been demonstrated that neural network models appear to be useful for performing spatial classification with multitype and noisy data, unequal reliabilities of data sources, non-smooth information distribution and non-linearity. While multilayer feedforward neural networks have been commonly used for such a purpose, it is pointed out in this chapter that other unidirectional and recurrent neural networks might be even more powerful for spatial classification and pattern analysis. In addition, neural-based models deriving from scale space theory have also been introduced here to perform spatial classification, especially clustering, by mimicking the blurring effect of our lateral retinal connections. It could turn out to be a powerful method for spatial classification and pattern analysis.

With all the enthusiasm and effort devoted to the application of the neural network approach to spatial classification, we would like to caution that neural network models are not the answer to everything. They deliver meaningful results only when the conditions, such as those discussed above, are correct. When the underlying assumptions stipulated by conventional models are met, neural network models often become superfluous.

Although evolutionary computations, for no obvious reason, are seldom employed to perform spatial classification, they have found applications in knowledge acquisition, generally captured as rules. In this chapter, two genetic-algorithm systems, SCION and GANGO, have been shown to give good performance in precise and fuzzy rule acquisition respectively. It has been demonstrated that genetic-algorithm-based models are effective in acquiring rules from highly irregular search space where there is no knowledge about its gradient, structure or parameters. Again evolutionary computations turn out to be superior only when these conditions are met. We have to understand our problems well in order to choose an appropriate model [conventional or evolutionary] for the task of rule extraction from learning examples. Parallel to evolutionary computations, neural network models have also been developed for rule acquisition. The radial basis function neural network discussed in this chapter is a typical example.

The potential of these two paradigms in solving difficult problems and in extending the analytical frameworks of spatial analysis in general and spatial classification and knowledge acquisition in particular have been echoed in some recent writings (Leung 1997; Openshaw and Openshaw 1997; Wilson 2000). Neural and evolutionary models appear to be able to do, with varying success, what conventional methods can do, but can also accomplish tasks which conventional models are unable to handle. There do however remain theoretical and implementation problems which need to be further investigated. To make neural and evolutionary computations convincing, useful and practical in the analysis of large scale spatial systems, we still have to face a lot of challenges in the years to come.

5 Cellular Dynamics: Modelling Urban Growth as a Spatial Epidemic

Michael Batty
Centre for Advanced Spatial Analysis (CASA), University College London

5.1 Defining Urban Growth as Sprawl

Contemporary urban growth consists of three interrelated problems of spatial dynamics: the decline of central or core cities which usually mark the historical origins of growth, the emergence of edge cities which both compete with and complement the functions of the core, and the rapid suburbanization of the periphery of cities – core and edge – which represent the spatially most extensive indicator of such growth. Our understanding of these growth processes is rudimentary, notwithstanding at least 50 years of sustained effort in their analysis. Our abilities to 'control and manage' such growth or 'sprawl' as it is colloquially and often pejoratively referred to, is virtually non-existent despite occasional but short lived successes through planning instruments such as green belts. The suburbanization of cities and methods for the control of such growth go back to the origins of cities themselves. Urban history reveals a succession of instruments used to separate the growing city from its suburbs. Documented examples refer to Ur in Sumeria, ancient Rome, to Elizabethan London, where edicts were in place to ensure the quality of life in the core city by restricting overbuilding and access (Morris 1979). However the concept of suburb has changed through history. Jackson (1985) sums this up quite cogently when he says: '... the suburb as a residential place is as old as civilization ... However, suburbanization as a process involving the systematic growth of fringe areas at a pace more rapid than that of core cities ... occurred first in the United States and Britain, where it can be dated from about 1815' (p. 130).

As Jackson (1985) implies, until the mid-20th century the problem of sprawl was coincident with the growth of the industrial city. This provided a focus for the transition from comparatively low density agricultural society to one dominated by much more intensive production and consumption in cities where economies of scale were achieved through very rapid population growth. The problem in the early 21st century is somewhat different. Production and consumption can now be spread out less intensively, although very specialized cores continue to grow at all levels of the urban hierarchy, and developments in transport and communications technologies enable populations to gain much wider access to facilities both physically and remotely. This image of a world composed entirely of cities and their suburbs has been anticipated for a long time. Almost one hundred years ago, Wells (1901, quoted in Hall 1988) wrote about a completely urbanized Britain loosely cemented by new forms of communication technologies, a prospect which is well on the way to realization, despite relatively modest population growth. In the United States, where population growth has been much greater, where there are far fewer constraints posed by the land supply, and where the incentives for suburban development are much clearer, the problem of sprawl is different again although no less significant (Nivola 1999).

Most of the focus of physical planning in western countries during the 20th century has been on ways of controlling urban growth, but the recent wave of economic and related forms of institutional deregulation have given the problem a new urgency. Traditional solutions such as reducing our ability to locate in suburban locations through selective taxation of travel, combined with incentives to develop residential and other activities nearer the core of old cities are being suggested once again. But such policies are doomed in that they ignore completely the structure of the modern spatial economy where the central city is now just one of many nodes within a complex sea of urbanization whose pricing and market structure almost defies understanding (Krugman 1993). More sensitive policies, particularly those being canvassed in North America, admit that such growth is going to take place and that it will be suburban, but more selective ways of letting it take place are being proposed. This is the idea of 'smart growth' which basically involves controlled or managed sprawl where balance is the watchword for developing communities in which there is much the same variety of opportunity in travel and recreation as there is in less controlled growth (Nivola 1999).

This brief commentary reveals our woefully inadequate understanding of urban growth processes, but it also suggests that there are many different types of process and thus many different varieties of urban sprawl. To develop a better understanding of such processes as a prelude to better classification, it is hard to escape the conclusion that we must return to first principles, and examine the growth process in terms of its fundamentals. In this chapter, we will begin this quest, abstracting urban growth to such a degree that we will concentrate on only the geometric properties posed by such growth in space and time. We will assume away the practical issues posed by preferences for travel, space and other amenities as well as policy issues reflected through taxation and the market. We will also assume away demand, and simply concentrate on how cities grow through the addition of space to their periphery, and through aging processes which determine the condition of their physical stock. As such, our models focus on generic processes which we assume are fundamental to growth and location; but these do not provide models that can be directly applied to real situations unless fiercely adapted to local conditions. In short, the tradition that we will adopt in this foray into urban growth modelling is aggregate and geometric, with little account of the kind of individual behaviours that define how populations react to prices and markets. Despite such simplicity and parsimony, we will argue that our models do provide insights into the dynamics of urban sprawl which are useful for intelligent discussion of the problem and its potential solution.

The key elements in the spatial dynamics of growth involve the *available space* within which growth takes place, and the *aging process* of development associated with that growth. A force or momentum for growth is required otherwise it will stop, but this usually relates to wider considerations which pertain to issues beyond the immediate spatial confines of the city in question. The overall level of growth will depend on the local demographic factors as well as the attraction of the city to new growth from outside. Changing preferences for the consumption of space will also affect growth, but all these can be linked to the amount of space available. Our simplest model of urban growth can be conceived quite coherently without recourse to any formal apparatus: imagine a city growing around a seed or core where the amount of new growth is proportional to the space available on its periphery, and that this space is unlimited in time, that is the city can keep on growing in this fashion in perpetuity. Then a wave of growth will move out from the core, as the city grows in such a way that space and time are coordinated. However, development has a limited life span. Buildings do not last forever and thus as the city grows, past waves of growth will decline as buildings are demolished. If we assume that new growth occurs immediately following this regular decline, new waves of growth appear and as the city grows larger, its growth dynamic becomes dominated by successive waves on a spectrum from new growth at its edge to redevelopment phased according to the life cycle of the city's historic development.

In so simple a form, this model does not exactly mirror reality but already we have an image of how some cities develop. Often when development comes to the end of its life, it is abandoned and waves of new growth, building on the old, do not take place. Central cities are depopulated and abandoned while the suburbs keep on growing. This is a process of succession without the kind of invasion that some cities experience when older neighbourhoods are rehabilitated and reoccupied. Of course, cities do not grow in perpetuity, and another element of our analysis will be to examine the effect of limiting the growth process on the waves of development that characterize this process. One element of our analysis does change the conventional analytic approach to suburbanization and sprawl. Often suburbanization and sprawl is assumed to embrace all the processes of change that are taking place outside the core of the city. In fact, this analysis will show that suburbanization occupies only a very small fraction of what is happening in the city at any one time, and that in mature cities, most change potentially comes from redevelopment and movement within the existing stock rather than new development. We will argue that herein lies one of the keys to solutions to the problems of sprawl.

The framework we adopt is based on an analogy between the process of converting land from non-urban to urban and the idea that an individual becomes infected with a disease and then recovers. In short, we will articulate land conversion as a process of infection and then recovery, following quite well-established ideas in the simulation of epidemics. Firstly, we introduce this method of modelling sprawl, showing its key elements for the non-spatial case and how it is linked to conventional growth processes based on exponential and logistic growth. We then generalize the model to a spatial context in which waves of growth are modelled as spatial diffusion, and show how redevelopment is intrinsic to this process. We implement the model as a cellular automata [CA] and illustrate its working for significantly different examples. This model enables us to vary key parameters controlling the spread of development, its timing, and the degree to which the growth process is subject to random influences. This provides a basis for us to classify different outcomes. Although our model is quite limited in import, we are able to show significantly different growth outcomes which provide a basis for classification. To progress this analysis to real world applications, we conclude by showing that this logic can be embedded within a wider CA framework based on the **DUEM** model (Batty, Xie and Sun 1999; Xie 1996), illustrating an idealized simulation of suburban growth using data for Ann Arbor, MI.

5.2 Growth as an Epidemic: Spatially Aggregate Models

Nivola (1999) makes the obvious point that cities '... can only grow in four directions, in, up, down and out ... ' with growth ' ... likely to follow the last of these paths overwhelmingly, particularly in advanced countries endowed with abundant usable territory' (p. 2). Such growth reflects a combination of influences and preferences: population growth which must meet the geometrical and resource constraints posed by the shape and technology of the city, the preference for newness, which is illustrated in higher demand and higher prices for new in contrast to old housing, the preference for lower densities which can best be met by development on the fringe of the city where more land is available, access to better environmental amenities which again suggest newer rather than older locations, and the desire to avoid traffic congestion. These elements are all reinforced in western countries by policies that provide a cornucopia of tax advantages to location in the suburbs. Empirically, the way western cities have grown over the last 200 years is entirely consistent with these forces. The notion that urban growth has advanced in waves of development outward from the central city or historic core is well established, as in Blumenfeld's (1954) characterization of the growth of Philadelphia, the same phenomena being implicit in much of the work on urban densities, as for example in Bussiere's (1972) work on Paris.

In treating the process of suburbanization as the cumulative growth of the city through additions to its periphery, we need to divide development at any time t into three constituent parts: development which is established [in this context surrounded by other development] which we call $P(t)$, new development $N(t)$ [which has just made the transition from undeveloped land], and available land $A(t)$ which drives the process of development in the first place. This process is one in which new development first makes the transition from undeveloped or available land, then once this development is no longer adjacent to any available land, it passes into a mature state as established development. When the land is first developed, this depletes the stock of available land. If this stock is limited, then eventually growth will cease, as the limit C of what can be brought onto the market is reached. In fact, in treating the problem as a spatial aggregate, we will make the assumption that the city and its hinterland have constrained capacity so that

$$P(t) + N(t) + A(t) = C, \forall t, t = 0,1,2, \ldots, T \tag{5.1}$$

From Equation (5.1), it is clear that any changes in each of the categories of development must also meet the constraint

$$\frac{dP(t)}{dt} + \frac{dN(t)}{dt} + \frac{dA(t)}{dt} = 0 \tag{5.2}$$

The mechanism in the model is one-directional: available land passes to new development which then passes to established development as $A(t) \rightarrow N(t) \rightarrow P(t)$. Usually the process begins with all land $A(0)$ set equal to the capacity $C - \varepsilon$, where the fraction ε is the seed of new development that starts the process of growth and transition, that is $N(0) = \varepsilon$. The pivotal change is in newly developed land from the addition of new development $dn(t)/dt$ and the transfer of such development to its established state, $dP(t)/dt$. Then

$$\frac{dN(t)}{dt} = \frac{dn(t)}{dt} - \frac{dP(t)}{dt} \tag{5.3}$$

We hypothesize, as in similar models of change, that new development associated with available land is a proportion α of the product of $N(t)\,A(t)$. We can consider $\alpha N(t)$ to be the proportion of new development that generates a unit of new development and the number of such units is defined by the available land $A(t)$. The transfer of new development to its mature state is also a proportion of γ of each unit of new development $N(t)$ and using these definitions equation (5.3) becomes

$$\frac{dN(t)}{dt} = \alpha N(t) A(t) - \gamma N(t) \tag{5.4}$$

Note that the term $\alpha N(t) A(t)$ is an interaction term that represents the essential mechanism of conversion in the model. The two other components of change follow directly as

$$\frac{dP(t)}{dt} = \gamma N(t) \text{ and} \tag{5.5}$$

$$\frac{dA(t)}{dt} = -\frac{dn(t)}{dt} = -\alpha N(t) A(t) \tag{5.6}$$

It is quite clear that these definitions meet the constraints posed by Equations (5.1) and (5.2). A simple block diagram of the relations is provided in Fig. 5.1 which shows that new development is an essential filter in the growth process which is articulated as the transition between undeveloped or available land and established development.

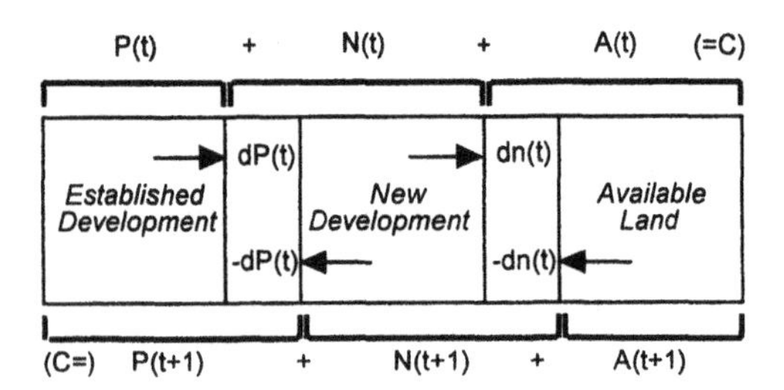

Fig. 5.1 An aggregate model of land development

Equations (5.4) to (5.6) are identical to those used to model epidemics. If we define new development $N(t)$ as infectives, available land $A(t)$ as susceptibles, and established development $P(t)$ as removals or infectives who have recovered from a disease, then the process can be likened to the infection of a susceptible population, with those infected being gradually removed from the population either through recovery or death. In this interpretation, the central mechanism of the model is the interaction between infectives and susceptibles and in this case, $\alpha N(t) A(t)$ has the immediate interpretation as the number of new infectives $dn(t)$ which is a proportion α of all the potential contacts between infectives and susceptibles. It is easy to guess what happens in this model since, as the number of susceptibles from the fixed pool declines, the number of new infectives also falls and ultimately the epidemic dies out. The model has been quite widely applied to various kinds of epidemic (see Murray 1993 for a good discussion), but it also has relevance to any problem which involves the transfer of resources from one state to another, a process that is widely used to model technological as well as political change (Banks 1994; Epstein 1997).

It is not possible to derive closed forms for the three differential equations (5.4) – (5.6) that define the process, although considerable algebraic analysis of the system is possible, and it is quite easy to show how the model behaves. In particular, it is clear that for growth in the developed land categories to take place at all, the initial amount of available land $A(0)$ must be greater than some threshold. If we write equation (5.4) as

$$\frac{dN(t)}{dt} = \alpha N(t)\left[A(t) - \rho\right] \tag{5.7}$$

where $\rho = \gamma/\alpha$, and noting that available land $A(t)$ must always reduce with time. As Equation (5.6) implies, for the number of infectives to increase, Equations (5.4) or (5.7) must always be positive. This means that $A(0) > \rho : \rho$ can be interpreted as a threshold which represents the minimum amount of available land necessary to get an epidemic going. Central to the generation of new development from available land is the ratio of the rate of transition of new to established development and the contact rate. In epidemic models, this is referred to as the 'relative removal rate', and such models are often called 'threshold-epidemic' models accordingly.

This point is made clear another way. If we take the ratio of equations (5.6) and (5.5) which we write as

$$\frac{dA(t)}{dP(t)} = -\frac{\alpha}{\gamma} A(t) = -\frac{A(t)}{\rho} \tag{5.8}$$

we can write an equation for land available in the steady state at $t = \infty$ as

$$A(\infty) = A(0)\exp\left[-P(\infty)/\rho\right] = A(0)\exp\left[-\frac{C - A(\infty)}{\rho}\right] \tag{5.9}$$

Equations such as (5.9) can be used to compute the steady state values of the three key components for different parameter values as well as the parameter values which are able to reproduce a steady state from fixed initial conditions (Murray 1993).

In Fig. 5.2, we show typical trajectories generated by this growth process where the capacity limit C is set as 1000. We start the process with $A(0) = C - 1$, $N(0) = 1$ and $P(0) = 0$. The parameters $\alpha = 0.000146$ and $\gamma = 0.03199$ give a relative removal rate $\rho = 219.12$ which implies that $A(0)$ must be greater than this value for growth [an epidemic] to take place. Solving equation (5.9) iteratively gives the steady state value for available land $A(\infty) = 10.94$ which means that not all available land is utilized for development. The fact that growth stops a little short of the capacity limit is simply a consequence of the parameter values α and γ. Nevertheless, what we see here is classic capacitated growth – logistic growth – but through a conversion process, i.e. a filter. Mature development grows according to the classic S-shaped curve, while available land declines as a mirror image of this. The structural diagram which illustrates the way the three components of the model are related in Fig. 5.1 suggests as much.

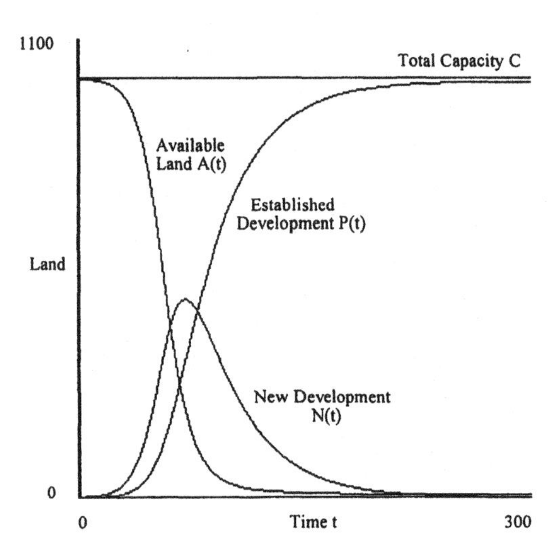

Fig. 5.2 Trajectories of development from the basic model

5.3 Simplifications and Extensions to the Aggregate Model

Some obvious simplifications need to be noted in order to ground the model in more conventional exponential and logistic growth processes. First assume that there is no transition from new to established development, that is $\gamma = 0$, meaning that new development becomes established immediately. In other words, the filter of new development no longer exists and we write the overall capacity constraint as $N(t) + A(t) = C$. Then the system of Equations in (5.4) to (5.6) collapses to

$$\frac{dN(t)}{dt} = \frac{dn(t)}{dt} = \alpha\, N(t)A(t), \quad \frac{dP(t)}{dt} = 0, \text{ and } \frac{dA(t)}{dt} = -\frac{dN(t)}{dt} \tag{5.10}$$

As the terms for $dN(t)$ and $dA(t)$ are symmetric, then we need only deal with one of these. Noting that $A(t) = C - N(t)$, we can write $dN(t)$ as

$$\frac{dN(t)}{dt} = \alpha N(t)[C - N(t)] = \alpha CN(t)\left[1 - \frac{N(t)}{C}\right] \tag{5.11}$$

Equation (5.9) is one form of the logistic equation which on integration yields

$$N(t) = C / \left\{1 + \left[\frac{C}{N(0)} - 1\right]\exp\left(-\alpha Ct\right)\right\}, \tag{5.12}$$

while the equation for $A(t)$ is the reverse of (5.11) which we illustrate in Fig. 5.3(a) for this growth process. This logistic is sometimes called the 'simple' epidemic model in contrast the 'general' model already introduced. The key insight here is that both the epidemic and logistic processes are capacitated, but the epidemic process involves a more complex transition from the undeveloped to the developed state. We can relax this model even further if we take off the capacity constraint, for it might be argued that there are many growth situations in cities that are not limited by available land. We will relax this assumption later in our more realistic spatial models, but here, when we remove the capacity constraint, a little algebraic manipulation of Equation (5.12) produces the exponential growth model: $N(t) = N(0)\exp(\alpha t)$.

There are other ways of adapting the general epidemic model to deal with the more realistic conditions of urban growth. If the capacity of the system $C(t)$ is made a function of developed land, that is

$$\frac{dC(t)}{dt} = \beta P(t) \tag{5.13}$$

where β is the rate of growth in capacity, then the change in available land – Equation (5.6) in the general aggregate model – becomes

$$\frac{dA(t)}{dt} = -\frac{dn(t)}{dt} + \frac{dC(t)}{dt} = -\alpha N(t)A(t) + \beta P(t) \tag{5.14}$$

The trajectories generated by this model are illustrated in Fig. 5.3(b). As in many such models, after some initial oscillation, the system produces growth at a constant rate. This is indicated by the convergence of $A(t)$ to a constant level, implying that all growth in available land is transferred directly into growth in new and thence established development.

A third variant of the basic model involves introducing a structural distinction between two varieties of available land: a suburban fringe defined by $F(t)$ which we can associate directly with new development, and a wider suburban belt or periphery which makes up the rest of the stock of undeveloped land. We will define this periphery as $A(t)$ and the capacity of the system is now composed of

$$P(t) + N(t) + F(t) + A(t) = C \tag{5.15}$$

The difference between this and the general model is that the land in the fringe is made a function of change in new development, thus incorporating the idea that the fringe depends on where new development is located, just as new development depends upon the fringe. This relationship is written as

$$\frac{dF(t)}{dt} = \mu \frac{dN(t)}{dt} \tag{5.16}$$

where μ is the transition parameter. Equations for new and established development are the same as previously – Equations (5.4) and (5.5) – but the change in available land is now set as

$$\frac{dA(t)}{dt} = \frac{dn(t)}{dt} - \frac{dF(t)}{dt}. \tag{5.17}$$

If we now add the change Equations (5.4), (5.5), (5.16) and (5.17), these sum to 0, thus ensuring that the total development in the system including fringe and peripheral land always equals the capacity of the system. The trajectories of this model are shown in Fig. 5.3(c).

This variant shows some of the difficulties in making *ad hoc* adaptations to this framework. Fig. 5.3 (c) illustrates that the problem with the model converging to a stable level of unused land is caused by the way the fringe interacts with this land and with new development. It is possible to predict the values of α, γ and μ which lead to different steady states, as we did with the general model, but the algebraic analysis becomes increasingly convoluted and is beyond the scope of this chapter. What this model does introduce is the notion that interaction between development and available land must be more sophisticated to ensure that the capacity constraint is only exercised when the system approaches this limit spatially. These models are not spatial, but to anticipate the models we develop below, we will resolve the problem by letting available land be generated locally, rather than imposing any global limit. Such a limit will in fact still operate, but only when development comes within 'sight' of this limit. This can only be accomplished when the model is specified spatially.

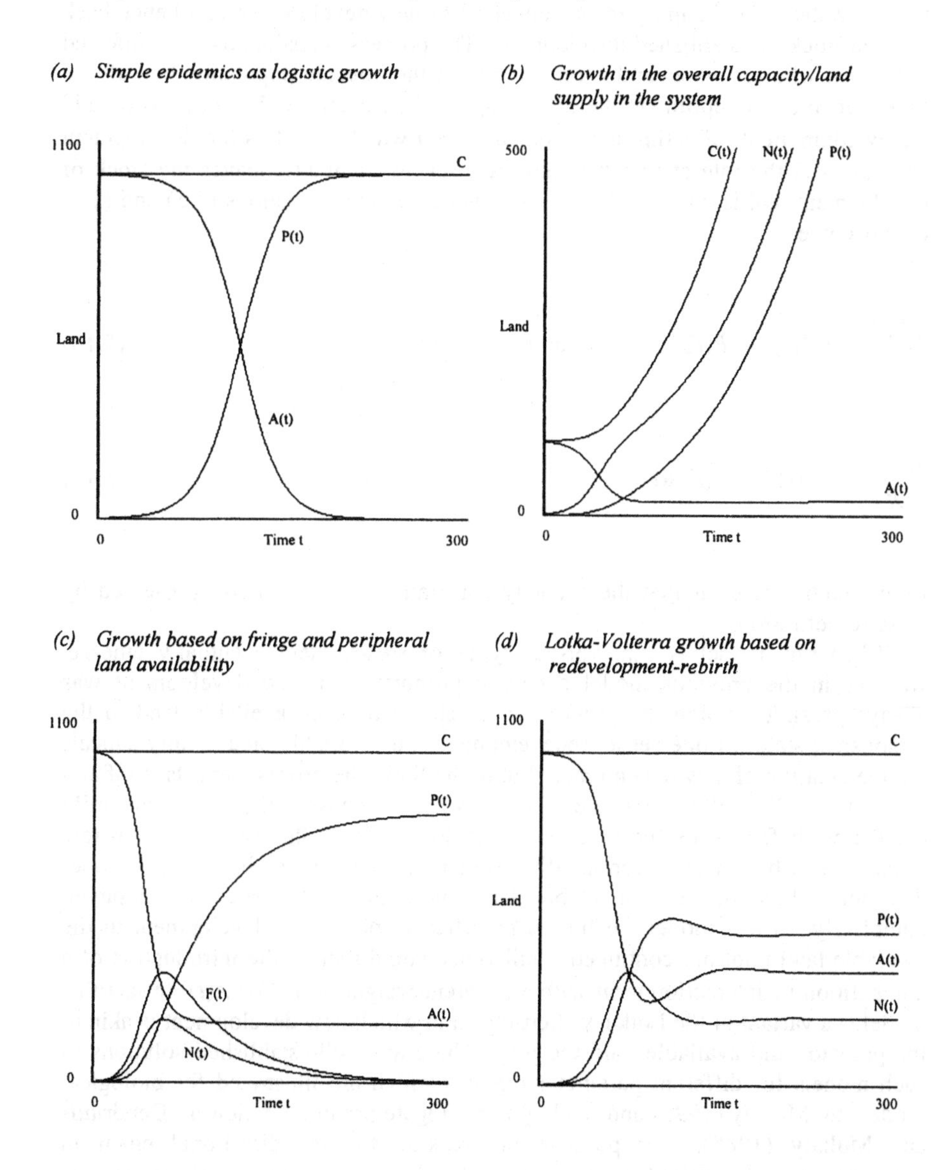

Fig. 5.3 Variants of the basic model

In all these models it is assumed that once development has been established, this state is irreversible. However development [like many diseases] has a life cycle which needs to be accounted for. Development ages physically and has an average life span, after which it is removed or demolished and new development takes it place. We model this as a process in which a proportion of established development comes up for redevelopment in each time period. This therefore adds to the available land supply to be converted to new development and hence back into the stock of established development. The process is analogous to an infected individual who recoveres and is removed from the infected population but, after a time, becomes susceptible to the disease again. To Equations (5.4) to (5.6) we add a new component reflecting development $\lambda P(t)$ which re-enters the development process. λ is the rate at which established development $P(t)$ leaves the stock of development and is converted back into available land. Equations (5.5) and (5.6) now become

$$\frac{dP(t)}{dt} = \gamma N(t) - \lambda P(t) \qquad \text{and} \qquad (5.18)$$

$$\frac{dA(t)}{dt} = \lambda P(t) - \alpha N(t) A(t) \qquad (5.19)$$

from which it is clear that the capacity constraint in (5.15) is still preserved by these re-definitions.

When this model is run, various types of steady state eventually emerge, whereas in the previous model a certain proportion of new development was always passing through the process of development from available land to the established state. In one sense, redevelopment into available land simply cancels out the creation of new development but in the limit, the process repeats itself at a stable level. With the parameter values we have chosen, there is some mild oscillation before stability is attained as Fig. 5.3(d) illustrates. To generate trajectories which would continually oscillate would require breaking the one-directional link between available land, new and established development completely. In this model the link is modified to pass back development to the available land pool but continued oscillations would require the introduction of a competition or interaction term with a different weight on $\alpha N(t) A(t)$. In fact this model is a variant of the Lotka-Volterra form in which new development is akin to the predator and available land the prey. There are well-established solutions to such models for different parameter regimes which are illustrated for biological models by Murray (1993) and for highly aggregate urban evolution by Dendrinos and Mullaly (1985). Our purpose in breaking the one-directional chain in development at this point to include renewal and redevelopment simply anticipates the way we will deal with these issues in the spatial models that are introduced in the next section.

Our last variant extends the direct logic of the epidemic model structure. As we have emphasized, the general model simulates a transfer of resources – available land – through a filter called new development, to an ultimate state – established development – in such a way that the available land supply always decreases, and development always increases. More than one filter can of course be placed between these two states of non-development and development if there is good reason. For example, the process might be conceived as one in which peripheral land $A(t)$ is first transferred to a semi-prepared state $F(t)$ [on the urban fringe perhaps], which is then newly developed as $N(t)$ and passed to its final state as established development $P(t)$. This in fact is more characteristic of the land development process, in that fringe land often remains for some years in this intermediate state before it is developed, seldom reverting back to available land first. The block diagram in Fig. 5.4 illustrates this logic which is exactly the same as in Fig. 5.1 but with $N(t)$ and $F(t)$ now both acting as filters. As in the general model, peripheral available land gradually declines but fringe land first increases [and then declines] with new development following the same process a little later, finally adding to the stock of established development.

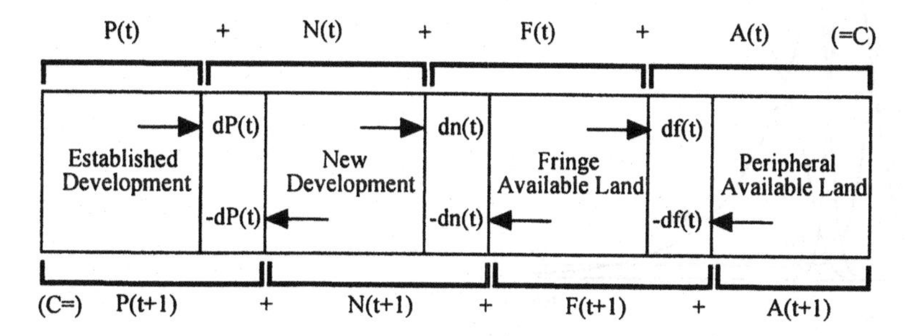

Fig. 5.4 Extending the aggregate model

The equations describing this process need to be made explicit and following earlier definitions, we will now present the entire set as

$$\frac{dP(t)}{dt} = \gamma N(t); \quad \frac{dn(t)}{dt} = \alpha N(t)A(t); \quad \frac{dN(t)}{dt} = \frac{dn(t)}{dt} - \frac{dP(t)}{dt}, \tag{5.20}$$

$$\frac{df(t)}{dt} = \sigma F(t)A(t); \quad \frac{dF(t)}{dt} = \frac{df(t)}{dt} - \frac{dn(t)}{dt}; \quad \frac{dA(t)}{dt} = -\frac{df(t)}{dt}, \tag{5.21}$$

where σ is the proportion of fringe land $F(t)$ interacting with available land on the periphery. The two key relations $dn(t)/dt$ and $df(t)/dt$ reflect the linking between the filters, newly developed land and fringe land determining new development, but fringe land and the periphery determining the amount of land that passes into the fringe. A typical example is illustrated in Fig. 5.5 which has the same steady state properties as the general model. Murray (1993) demonstrates ways in which this variant might be further developed. More filters could be added in the same way and with the same recurrence of patterns. It is possible to extend this more general model to incorporate the other variants developed in this section, particularly the one involving redevelopment. However, at this point the basic logic of showing how land can be transferred through a process which is similar to the way epidemic waves sweep through a population has been illustrated and we will now generalize the model to a spatial context, first by considering how such development might diffuse across space.

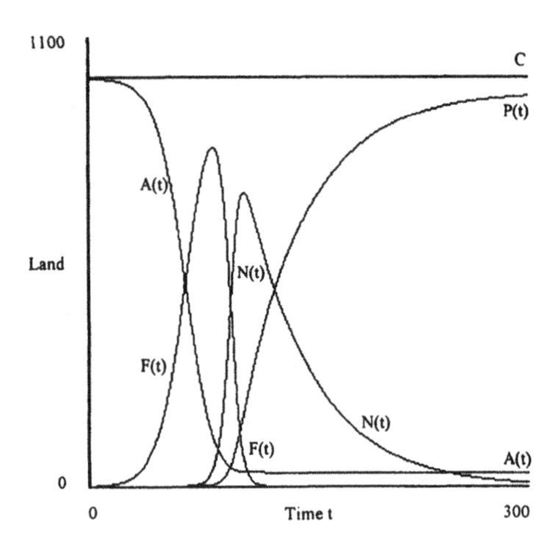

Fig. 5.5 Trajectories of mature, new, fringe and peripheral land development

5.4 Growth as Spatial Diffusion: Spatially Disaggregate Models

The translation of the general model to spatial coordinates x,y involves several changes in interpretation in that each cell or coordinate location has its own model while the interactions between locations are usually modelled through diffusion. First, land available across all locations does not formally enter the model framework as a constraint. Second, available land and new development are generated locally through diffusion to adjacent or nearby locations; that is land which is available for development at location x, y generates more land for development at locations adjacent to x, y. The same process can be articulated for new development which generates further new development around it. We now define the three land components as $A(x,y,t)$ – available land, $N(x,y,t)$ – new development, and $P(x,y,t)$ – established development, and state the model as follows. Change in new development is specified as

$$\frac{\partial N(x,y,t)}{\partial t} = \alpha N(x,y,t)A(x,y,t) - \gamma N(x,y,t) + D_N \nabla^2 N(x,y,t) \tag{5.22}$$

where D_N is a diffusion coefficient associated with new development. Change in available land is defined as

$$\frac{\partial A(x,y,t)}{\partial t} = -\alpha N(x,y,t)A(x,y,t) + D_A \nabla^2 A(x,y,t) \tag{5.23}$$

where D_A is a diffusion coefficient associated with available land. New development passes to established development in a similar manner to that used in the aggregate model as

$$\frac{\partial P(x,y,t)}{\partial t} = \gamma N(x,y,t) \tag{5.24}$$

It is not meaningful to sum these change equations across x,y since all they show is the fact that available land and new development diffuse across the space according to how much land there is locally around new development. Any global constraints on the capacity of the whole system are thus operated externally, as we will see in the discrete version of this model which we operationalize below.

This model has been less widely applied than might be imagined, notwithstanding some seminal contributions to the spatial diffusion of epidemics. It is mathematically complex in that it generates waves of development that diffuse across space but are bounded by the model's parameter values and diffusion coefficients. Richardson (1941) appears to have been the first to propose a similar structure for the movement of population, but applications to the urban domain have been restricted to theoretical illustrations based on spatial generalizations of the Lotka-Volterra framework with diffusion (Bracken and Tuckwell 1992; Zhang 1988). Most applications however have been to epidemics in human and animal populations, particularly measles, rabies, and the plague (Cliff et al. 1981; Cliff, Haggett and Smallman-Raynor 1998; Murray 1987; Noble 1974; Raggett 1982). There are also suggestions that the model might be applicable to social phenomena such as revolutions (Batty 1999; Epstein 1997).

There is a large literature on the geographical diffusion of settlements (Morrill 1968), but none of these examples have been embedded in the transition processes that we examine here.

The model in Equations (5.22) to (5.24) is not in a suitable form for applications to urban growth, as the processes of diffusion that characterize development are hardly likely to involve the kind of diffusion implied above. However, the way this model emphasizes local growth through diffusion is characteristic of the way an urban fringe develops and thus the framework can be easily adapted to suburbanization. Urban areas grow around their edges primarily due the demand for new space which is translated through the usual process into new development. New land becomes available adjacent to new development largely because of the existence of that development and thus the relevant diffusion is of new development determining additional land supply in its local neighbourhood. A more appropriate model form is based on

$$\frac{\partial N(x,y,t)}{\partial t} = \alpha N(x,y,t) A(x,y,t) - \gamma N(x,y,t) \quad \text{and} \tag{5.25}$$

$$\frac{\partial A(x,y,t)}{\partial t} = -\alpha N(x,y,t) A(x,y,t) + D_A \nabla^2 N(x,y,t) \tag{5.26}$$

where the spatial coupling is now through the diffusion term in Equation (5.26). Equation (5.24) is unchanged. In this model, land is made available through the existence of new development although new development still depends functionally upon the existence of prior new development and available land.

In the model we will build, there is still a global capacity constraint but as land and development are local operations, then the global constraint only becomes relevant at the point in time when the system approaches this limit. Available land is not affected by this limit until the diffusion reaches the local neighbourhood of the limit and then growth stops rather quickly. Therefore most development simulated by this model is locally motivated and occurs without reference to the overall capacity of the system. It is this that makes the spatial model considerably more appropriate, even in theoretical terms, than the aggregate models of the previous sections. Furthermore, in the model based on Equations (5.24) to (5.26), we will also incorporate the development cycle initiating redevelopment and renewal in the same way we introduced earlier. In the continuous version of the spatial model, such time delays have been studied in some depth and it is possible to simulate waves of development across space and time which are set off by such processes (Murray 1993; Zhang 1988). However in the computable form we will develop here, our analysis is more empirical and exploratory. To take this further, we must now move from a continuous to discrete structure in formulating operations analogous to those in Equations (5.24) to (5.26).

5.5 A Computable Structure Based on Cellular Automata

Let us assume a cellular space based on a regular tessellation of grid squares whose referent is given by the coordinate pair x,y. In analogy to the continuous model in (5.24) to (5.26), we define available land as A_{xyt}, new development as N_{xyt}, and established development as P_{xyt}. Each cell can be either filled or empty with respect to the activity in question implying equal densities of development per cell. To indicate cell occupancy, we set the relevant activity equal to 1, that is $A_{xyt} = 1$ or $N_{xyt} = 1$ or $P_{xyt} = 1$, such that no cell can have more than one activity, that is $A_{xyt} + N_{xyt} + P_{xyt} = 1$. We now consider t a time period. In any such interval the development process consists of three key transitions based on: (5.1) the diffusion effect which adds to the available land supply in the neighbourhood of each unit of new development equivalent to the term $D_A\nabla^2 N(x,y,t)$, (5.2) the transition from available land to new development equivalent to $\alpha_N(x,y,t)\,A\,(x,y,t)$, and (5.3) the transition of new development to established development equivalent to $\gamma_N(x,y,t)$. Each of these three changes form the components of the continuous model Equations (5.24) to (5.26), which we shall deal with in turn.

The diffusion effect is computed around each cell of new development by making adjacent cells available for development:

$$\left.\begin{array}{l} \text{if} \quad N_{xyt} = 1 \quad \text{then} \quad A_{ijt+1} = 1 \quad \text{where } i = x \pm 1, j = y \pm 1 \\ \qquad\qquad\qquad \text{unless} \quad N_{ijt} = 1 \text{ or } P_{ijt} = 1 \end{array}\right\} \tag{5.27}$$

Available land can thus come onto the market in any of the eight cells surrounding the new development in question, unless the cell is already developed. These cells lie at the eight points of the compass in a regular grid and define the so-called 'Moore' neighbourhood which is the most widely used in cellular automata modelling (Toffoli and Margolus 1987). In this sense, therefore, we can consider the model to be a variant of CA although this is entirely dependent upon defining the neighbourhood in this local way. The second transition is from available land to new development:

$$\text{if} \quad A_{xyt} = 1 \quad \text{then} \quad N_{xyt} = 1 \text{ and } A_{xyt+1} = 0 \tag{5.28}$$

where the term $\alpha N\,(x,y,t)\;A\;(x,y,t)$ is translated into this simple unweighted transfer. The last transition is from new to established development. In essence, we assume that when a cell of new development is no longer adjacent to any available land, then it is regarded as having entered the mature cycle. This is encoded in the model as:

$$\text{if} \quad \sum_{i=-1,+1} \sum_{j=-1,+1} \left[N_{x+i,y+j,t} + P_{x+i,y+j,t} \right] = 8 \quad \text{then} \quad P_{xyt} = 1 \text{ and } N_{xyt} = 0 \tag{5.29}$$

It is a very simple process for, in this form, available land and new development are separated by one time period. From a single seed, the process produces a wave of available land advancing one time period ahead of new development while, behind the wave, new development becomes established. On a grid, this can be visualized as a square band of land one cell wide advancing in front of a band of new development one cell wide, which is then converted to established development – the lag separating each component being one time period. When the edge of the system is approached, growth immediately stops for the capacity limit is only recognized locally through Equation (5.27). Of course it is possible to vary this process by changing the size of the neighbourhoods and incorporating differential time lags, but a more important issue is to break the spatial symmetry by introducing noise into the system.

It is much more realistic to assume that new development is generated from available land according to a certain probability, thus reflecting the fact that different developers of different cells vary in the way they finance development and react to the market. Such an extension would clearly break spatial symmetry, producing development clusters with irregular edges much more characteristic of real cities. A straightforward modification to Equation (5.28) achieves this:

$$\text{if} \quad A_{xyt} = 1 \text{ and } random(\Lambda_{xy}) > \Phi \quad \text{then} \quad N_{xyt} = 1 \text{ and } A_{xyt+1} = 0 \tag{5.30}$$

where *random* (Λ_{xy}) is a random number from a predefined range and Φ is a threshold above which land is converted to new development. If for example, Φ = 950 and the range is from 0 to 999, then this implies that if a random number drawn from the range is less than 950, then development would not take place. This in turn implies that about 95 times out of 100, new development would not take place on the set of cells which constitute available land [in the time period in question].

Aggregate growth trajectories equivalent to those in the aggregate models presented previously can be computed as

$$A_t = \sum_{x,y} A_{xyt}, \quad N_t = \sum_{x,y} N_{xyt}, \quad \text{and} \quad P_t = \sum_{x,y} P_{xyt} \tag{5.31}$$

When these are plotted for development around a single seed placed at the centre of a square bounded grid, A_t and N_t produce waves which are coordinated in space and time, rising exponentially in size at first and then decreasing rapidly as the boundary of the system is reached. This is also reflected in the path of established development P_t which reflects logistic growth although the logistic effect only kicks in when new development reaches the nearest boundary cells in the N-S-E-W directions, illustrating that capacitated growth is only a feature of the simulation once the boundary has been sensed. This is somewhat different from the aggregate model above where this capacity is always accounted for through the interaction between development and available land.

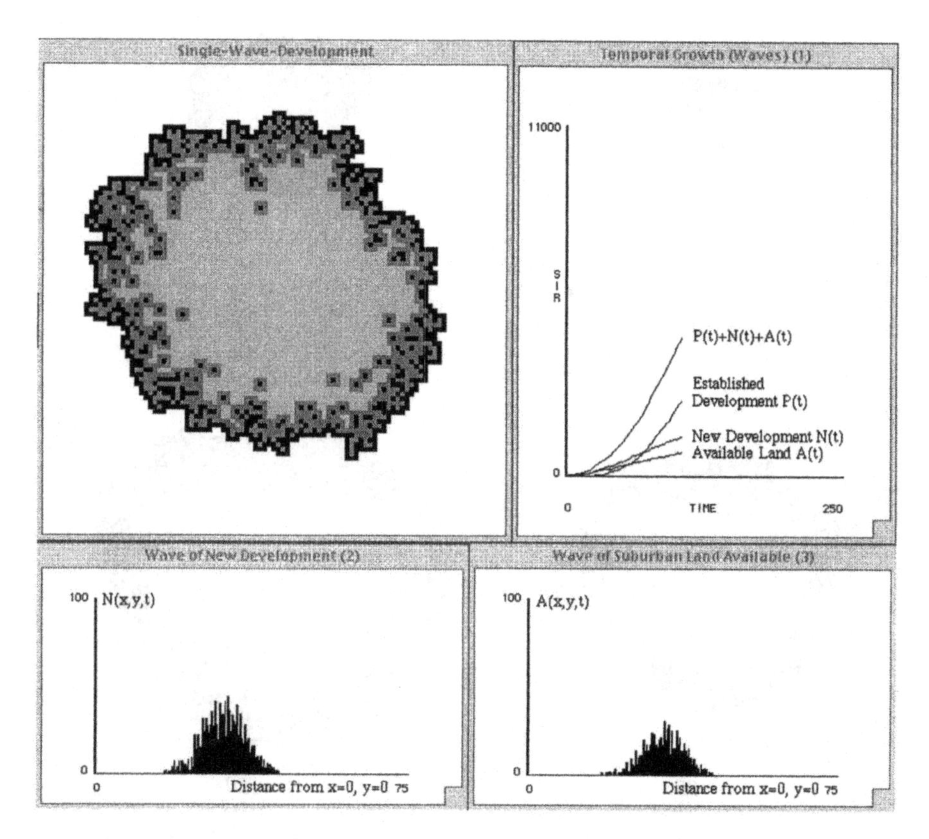

Fig. 5.6 Spatial diffusion from a single seed: Growth and morphology

Our first example plants a seed of new development at the centre $x = 0, y = 0$ of a grid whose dimensions are 101 x 101. We have set the development threshold at $\Phi = 900$ which implies on average only 10 percent of our transfers from available land to new development are successful. However, this simply delays the development process and does not widen the fringe of land available for development very much. The key feature of this probabilistic process is that it breaks the spatial symmetry. There are no other parameters in this model, and in Fig. 5.6 we show the development of the growing cluster at $t = 105$. Light grey is established development, mid grey new development, and black available land.

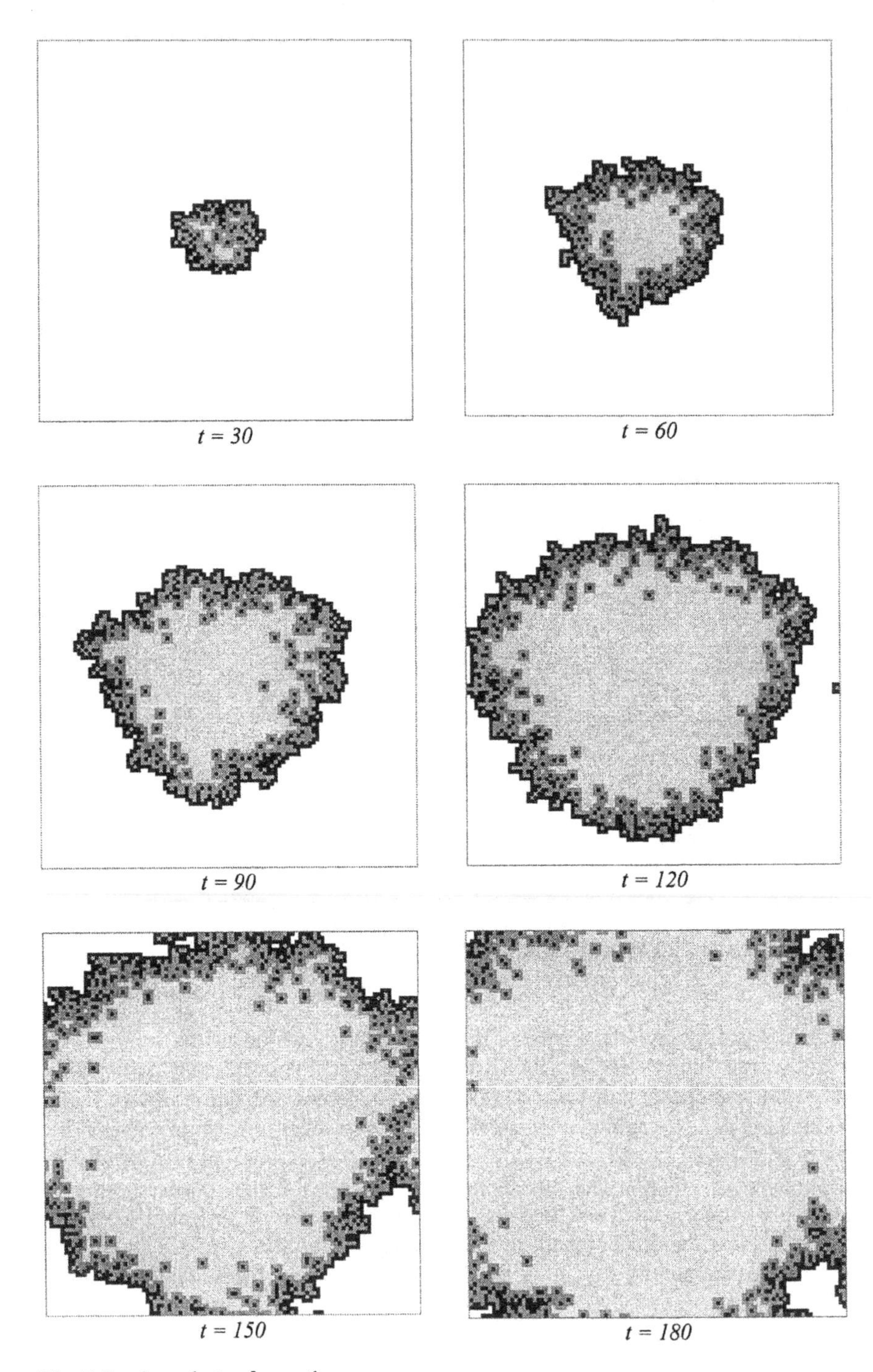

Fig. 5.7 Snapshots of growth

The cluster is characteristic, at least in terms of its boundary, of compact urban development. However, this level of compactness – development without holes – is unusual in western cities and later we will relax this assumption. The three other windows alongside the cluster show the distribution of new development and available land over space [measured from the centre $x = 0$, $y = 0$ up to a distance of 75 units], and the total activities A_t, N_t and P_t together with the total land affected by the development process $C_t = A_t + N_t + P_t$. The two windows which are associated with the spatial distribution of new development and available land show waves of land development. These waves move outward from the centre as the simulation proceeds and, in the software used, these are updated throughout the simulation. The waves grow in size as the cluster grows, but as the cluster approaches capacity, the waves gradually decline.

Each of the windows is coordinated within the software used, and Fig. 5.6 is thus a single picture from an animation across 250 time periods, illustrating the way the system builds up and eventually reaches a steady state in which all development is established and growth ceases. Six snapshots from this process are shown in Fig. 5.7, where it is clear that ultimately the cluster is composed only of established development. The overall growth trajectories are shown in Fig. 5.8 after the simulation reaches $t = 250$. In one sense, the waves mirror those of the aggregate model in Fig. 5.2. It also shows the space remaining, $R_t = C - C_t$, which mirrors the distorted S-shaped curves associated with P_t and C_t. Note that, if required, we can also compute all the change variables associated with these stocks, but these do not show any surprises and can easily be visualized by considering the first derivatives associated with the growth paths in Fig. 5.8.

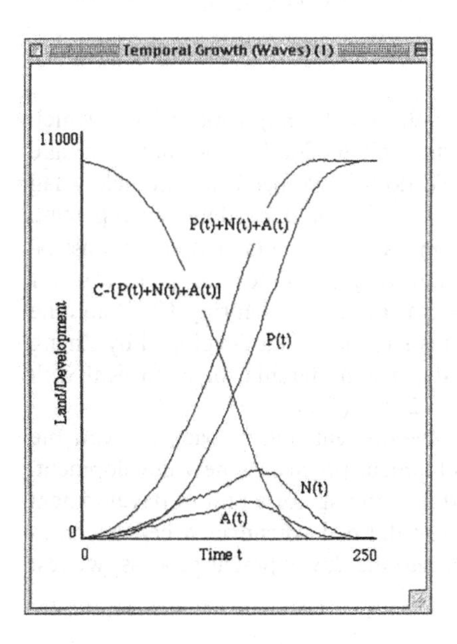

Fig. 5.8 Waves of development within logistic growth

We should also note that spatial epidemic models have been conceived in this simple way before. The models presented by Durrett (1995) have many similar characteristics to those presented in this section, although Durrett's purpose was to lay bare statistical assumptions rather than engage in actual simulation. Finally, it should be noted that some fractal models of urban growth have been based on diffusion and employ the same kinds of local cellular operators (Frankhauser 1994).

5.6 The Dynamics of Urban Regeneration

In the aggregate model, we indicated how established development might revert to available land, thus initiating the development process once again for the locations in question. Putting established development back into the development process is in fact one of the key processes of urban redevelopment or regeneration. To derive a continuous spatial equivalent, we modify Equations (5.24) and (5.26) as

$$\frac{\partial P(x,y,t)}{\partial t} = \gamma N(x,y,t) - \lambda P(x,y,t) \tag{5.32}$$

$$\frac{\partial A(x,y,t)}{\partial t} = -\alpha N(x,y,t) A(x,y,t) + D_A \nabla^2 N(x,y,t) + \lambda P(x,y,t) \tag{5.33}$$

where $\lambda P(x,y,t)$ is the proportion λ of established development $P(x,y,t)$ which comes back onto the market as available land. This model does not however take account of the life-cycle of development. To do this we need to introduce a lag consistent with the age of the development. This could be done by replacing $P(x,y,t)$ with $P(x,y,t-\tau)$ where τ is an age limit after which development must be redeveloped. Models such as these have been developed in the continuous case for the study of the spatial spread of disease such as rabies (see Murray 1993) and the form in Equations (5.25), (5.32), and (5.33) is similar to that developed by Zhang (1988) for urban development. As previously, it is preferable for us to deal with the discrete CA form and extend Equations (5.27) to (5.29).

We need to consider how established development reverts back to available land and thence once again enters the development process as new development. To do this at each time period, we need to define the age of a unit of development – the age of each cell x,y – as B_{xy}, which is defined whenever a cell becomes newly developed. To put development back into the development process, we test the following:

$$\text{if} \quad t - B_{xy} = \tau \quad \text{then} \quad A_{xyt} = 1 \ \text{and} \ P_{xyt+1} = 0 \tag{5.34}$$

where τ is the age threshold of the development at which it must be redeveloped or renewed through demolishing the structure, clearing the land, and placing the land in question back into the pool of available land to be considered for new development during the next time period. This rule could be randomized on the assumption that there is variation in this decision but we have not done this here as it only adds to the number of possibilities which might be simulated.

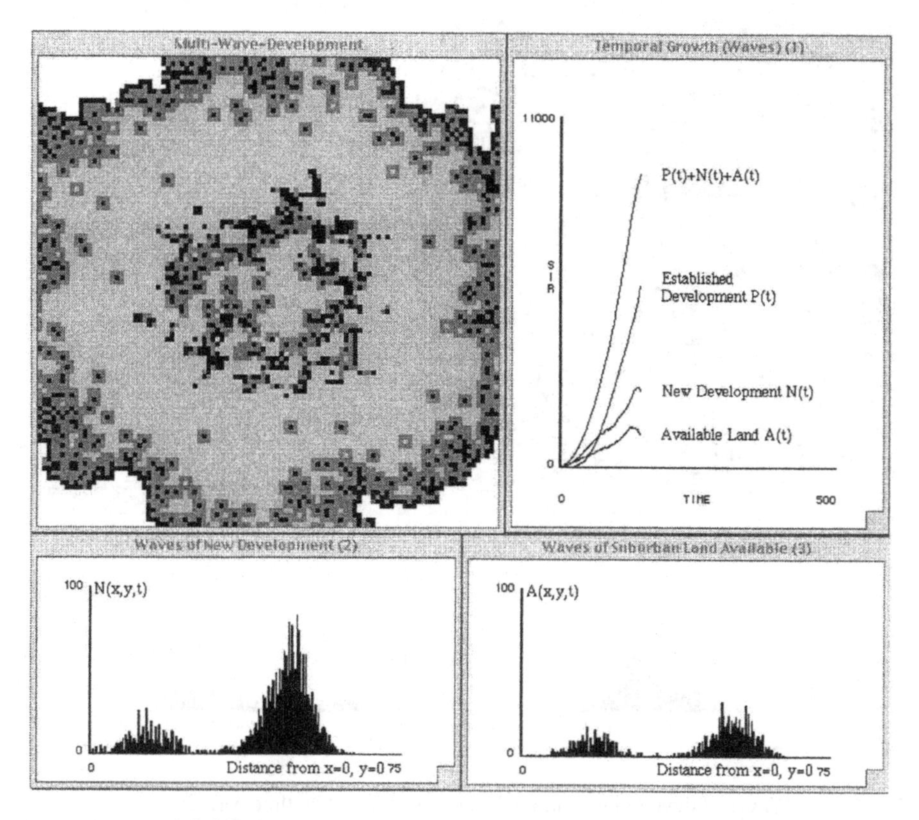

Fig. 5.9 Spatial diffusion with waves of redevelopment in every 90th time period

Using the same cellular space as previously (in Fig. 5.6), and now setting the threshold for redevelopment as $\tau = 90$, we show the morphology and waves of development at $t = 146$ in Fig. 5.9. It is immediately apparent from each of the four windows, that a new travelling wave of development from redevelopment, begins at $t = 90$. The initial wave moves outward from the seed of the cluster and this can be identified as the first – most right-hand – wave in the new development and available land windows, followed by the second wave some 90 time periods behind. This second wave also embodies the randomness of new development produced by the probabilistic land conversion process and although this second wave is associated with the exact times at which the first wave reaches the redevelopment threshold, this is further randomized by the conversion once again to new development. The second waves are thus less sharp than the first. Were we to randomize the time at which the age threshold for redevelopment were activated, this would spread these waves out even further.

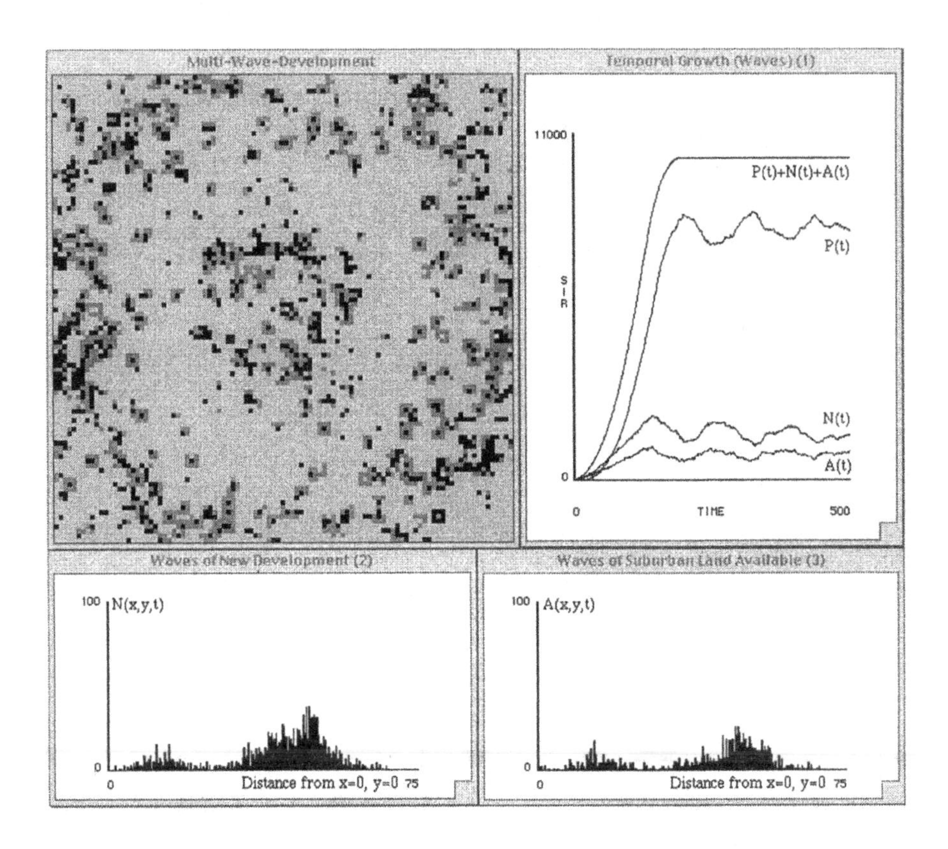

Fig. 5.10 Waves of development and redevelopment over 500 time periods

We now show what happens when this process is simulated over 500, then 2000 time periods. Every time a site is (re)developed, more noise is introduced into the actual time it passes from available land to new development. Over time, this means that two sites which were originally developed at the same time, might find themselves being redeveloped at times which are more and more distant from one another. In short, although initially the two sites were part of the same wave of growth, as they are redeveloped they are more likely to be in different waves; or in fact, the waves themselves are more likely to become less distinct. More and more noise is thus introduced into the spatial cluster each time sites are redeveloped and thus the initial spatial-temporal structure which might be very distinct, begins to break down. In Fig. 5.10, we show the pattern of development of the growing cluster after 500 time periods during which redevelopment of all sites will have taken place at least 4, probably 5 times. Because the age threshold is 90 and given the speed of growth in this model, it is likely that only two waves of development are distinct at any one time in a space of 101 x 101 cells. In Fig. 5.10, these two waves are much less sharp for new development and available land than after 146 time periods as in Fig. 5.9. We can see in Fig. 5.10 how these waves are reflected in aggregate trajectories. Because the system is capacitated, available land and new development, as well as established development, oscillate on a cycle of 90 years. This reflects the fact that a fully developed system never quite results due to the fact that, in the redevelopment process, a proportion of land always remains undeveloped. Although we cannot demonstrate this here, the simulation is highly sensitive to the age threshold and the system becomes much more volatile as this is decreased.

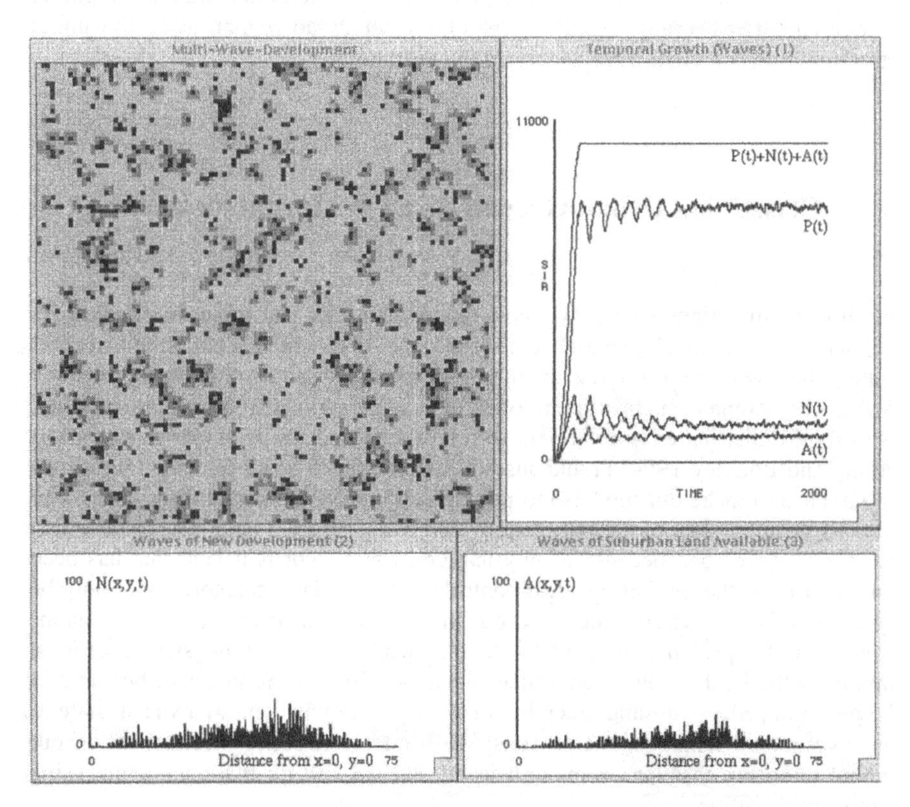

Fig. 5.11 Disappearing spatial and temporal structure through successive mixing

We continue the simulation to 2000 time periods in Fig. 5.11. By this time, the local waves across space and time and the spatial pattern itself have become quite unstructured. This is clearly evident in the aggregate trajectories where the system has more or less converged to a steady state in which the level of redevelopment is almost constant during each time period. Note that this situation has been reached after 20 or more cycles of redevelopment, which only begin when the first development reaches t = 90. The oscillations in fact become more and more frequent and their amplitude dampens considerably. The travelling waves of new development and available land have become relatively unstructured, as they have spread out in space and time, while the morphology of the cluster shows that the original concentric pattern of growth has all but disappeared. If we had added a random component to the time when the age threshold activated redevelopment, then these waves and patterns would be even less structured because another level of mixing would have occurred. This kind of randomness is in fact characteristic of real cities, and the kind of mixing of new development and sites becoming available for redevelopment that marks Fig. 5.11 virtually wipes out the original structure, which emerged initially from development around a single site. In terms of this model, there is no clue from Fig. 5.11 to how the initial structure might be recovered. In real cities, despite such mixing, clues do remain as to the initial structures and rules that have generated the morphology.

5.7 Classifying Urban Growth through Morphology

In all our simulations so far, we have ignored the fact that urban morphology is peppered with undeveloped and semi-developed land, which often remains in this state in the presence of rapid growth. In fact, just as urban form is fractal, then the 'holes' that comprise such structure often display an equivalent degree of fractality, with 'holes' distributed similarly in terms of size and shape across many scales (Batty and Longley 1994; Frankhauser 1994). We need to introduce a fourth type of land use to make our models morphologically more realistic in these terms, and thus we define a state of vacancy for any cell as V_{xyt}. Vacant land is not land which cannot be developed because of physical constraints, nor is it land that has been withheld from the market by legal controls, for such land categories can only be accounted for by external inputs to our models. This land is that which remains vacant in the presence of growth due to behavioural, not physical decisions, involving the land development process. It may be the wrong shape, or be part of a longer term process of land assembly or it may simply be kept in a vacant state as an investment. Whatever the reasons, we need to introduce mechanisms in our model to ensure that land remains vacant, and we will do this once again using simple probabilities.

Because our model has slowly become more and more complex, we will restate all the model equations which now involve vacant land, noting that any cell can only be in one use; this is equivalent to ensuring that $V_{xyt} + A_{xyt} + N_{xyt} + P_{xyt} = 1$, where $V_{xyt} = 1$ when it is vacant, 0 otherwise. Let us assume again a simulation which begins with the central cell in the system as new development, that is $N_{000} = 1$, and $N_{xy0} = 0, x \neq 0, y \neq 0$ with $A_{xy0} = 0$, $V_{xy0} = 0$, $P_{xy0} = 0, \forall xy$. We will present a typical simulation of the model in the order in which the various model operations take place for any time period t. First the age threshold is checked and if the development has reached the age at which it must be renewed it is put back into the pool of available land

$$\text{if} \quad t - B_{xy} = \tau \quad \text{then} \quad A_{xyt+1} = 1 \text{ and } P_{xyt+1} = 0 \tag{5.34}$$

Then the diffusion effect which determines whether or not available land [around new development] comes onto the market is effected

$$\left.\begin{array}{l} \text{if} \quad N_{xyt} = 1 \quad \text{then} \quad A_{ijt+1} = 1 \text{ where } i = x \pm 1, j = y \pm 1 \\ \qquad\qquad\qquad \text{unless } N_{ijt} = 1 \text{ or } P_{ijt} = 1 \end{array}\right\} \tag{5.27}$$

The new vacancy criterion is then checked. If land is available and/or already vacant, it becomes or remains vacant if a probability threshold Γ is exceeded:

$$\text{if} \quad A_{xyt} = 1 \text{ or } V_{xyt} = 1 \text{ and } random(\Lambda_{xy}) > \Gamma \quad \text{then} \quad V_{xyt+1} = 1 \tag{5.35}$$

The neighbourhood constraint which determines whether new development is transferred to mature development is now augmented with the vacant state as

$$\begin{array}{l} \text{if} \quad \sum\limits_{i=-1,+1} \sum\limits_{j=-1,+1} \left[N_{x+i,y+j,t} + P_{x+i,y+j,t} + V_{x+i,y+j,t} \right] = 8 \\ \qquad\qquad\qquad \text{then} \quad P_{xyt} = 1 \text{ and } N_{xyt} = 0 \end{array} \tag{5.36}$$

Finally the transition from available to new development is made, noting that at this stage the age of the development is determined:

$$\begin{array}{l} \text{if} \quad A_{xyt} = 1 \text{ and } random(\Lambda_{xy}) > \Phi \\ \qquad\qquad \text{then} \quad N_{xyt} = 1,\ B_{xy} = t,\ V_{xyt} = 0, \text{ and } A_{xyt+1} = 0 \end{array} \tag{5.37}$$

In this model, there are only three parameters τ, Γ and Φ, and one of these, τ, controls the redevelopment process. In the experiments we report in this section, we will set τ to a large number to ensure that redevelopment has no effect, and concentrate on the morphologies that result when we vary the values of the vacancy and spread parameters, Γ and Φ respectively.

The effect of the spread parameter Φ is to vary the speed of development. In the basic model with redevelopment and vacant land, all that Φ does is to slow down development as the value of the parameter is increased. Because there is no vacancy possible in this model and all land is ultimately developed, when Φ is large and near 1000, sites in the neighbourhood of new development have a low chance of being developed. But as the growth process continues indefinitely, until the capacity of the system is reached, at some point they will get developed. This makes a difference to the edge morphology of the system. When Φ is small, the system develops rapidly with little irregularity to the urban fringe. However, when we introduce the vacancy parameter Γ, if development is proceeding slowly with a high value of Φ, then land which is not developed at any time period, but is within the available land pool, has a continual chance of falling vacant, whereas with faster development this chance is lower. In short, it would appear *a priori* that development is increasingly unlikely as both the vacancy and spread thresholds are increased.

We have modelled the impact of a range of values for the vacancy and spread parameters, and for each of the clusters which develop, we have measured the final proportion of vacant land when the growth process is terminated. This occurs when no more available land is generated from new development. In Fig. 5.12 we plot changes in the proportion of vacant land associated with a range of vacancy parameters from 0 to 700 for two growth situations: very slow growth with $\Phi = 975$ and very fast growth with $\Phi = 50$. The functions generated are very similar in shape but displaced from one another with respect to the values of the vacancy parameter. What this graph shows is that for small values of Γ and Φ, almost complete clusters are generated with a low proportion of vacant sites [less than 10 percent]. However for the large value of Φ – slow speed of growth – the vacant sites increase very quickly, and once the vacancy parameter reaches around 400, then the growth of the cluster is halted around its seed. This is in marked contrast to the low value of Φ – fast speed – where vacant sites in the growing cluster are much harder to generate and retain. But by the time the vacancy parameter has reached 700 or so, the combination of this with the growth speed halts development before it gets a chance to start.

It is however the morphologies that are of real interest here, since the combination of the two parameters does give rise to fractal growth under most regimes. In Fig. 5.13, we have graphed a series of relationships between the percentage of vacant sites in the steady state and the vacancy parameter, the family of curves associated with different speeds of growth. To evaluate the resultant morphologies, we have sampled the space showing various forms for 10 sets of parameter values. In fact the shapes that we show are not those of the steady state, for we need to give an impression of how the clusters develop. In the steady state, as the system is capacitated, spatial irregularity is harder to observe; the system is bounded by the square grid, although it would be possible to colour code the development to provide a general impression of this irregularity. What we have done is to stop the growth as the cluster begins to approach the boundaries of the grid. As the grid is a torus, the usual default of graphic computer screens, development does wrap around, but the clusters shown give an immediate sense of the size and distribution of vacant sites under different combinations of threshold values.

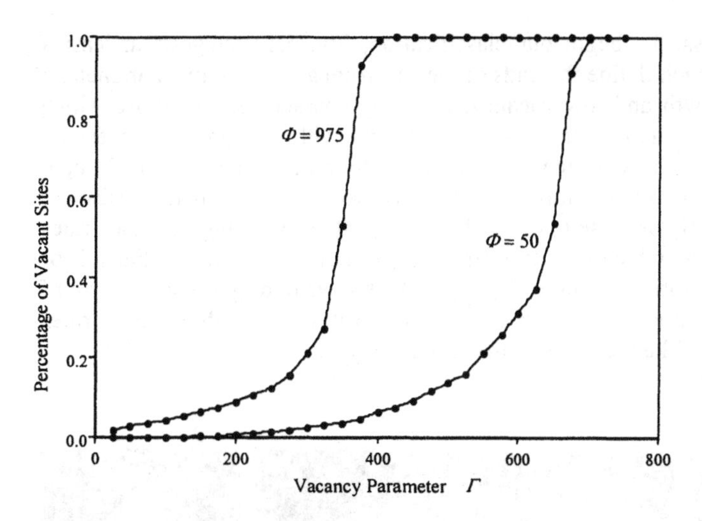

Fig. 5.12 The parameter space $\{\Phi, \Gamma\}$ associated with the proportion of vacant land

From the static snapshots of growth that we provide in Fig. 5.13, we are not able to provide a sense of how these clusters develop. In the sparser examples with higher vacancy rates and faster speeds, development often proceeds dendritically with the tips of the dendrites being the focus of development, although because the development wraps around in the closed spaces we have used, these dendrites soon disappear. In some measure, the pattern of development is not unlike that of a forest fire, with tips of the growing cluster being ignited as available land comes onto the market and then being turned into new development. With high levels of vacancy, many of these tips do not develop. If you look closely at the pictures in Fig. 5.13, you will see the tips composed of available land and new development which drive the process either to extinction or to complete occupation of the space. These examples show the power of this framework to simulate relatively realistic forms, despite the fact that we have not used varying neighbourhood sizes, multiple seeds, transportation networks, and all the other features that clearly have an influence on urban form.

We have not yet undertaken a systematic analysis of the fractal dimensions of the vacant sites in the samples in Fig. 5.13, but casual calculations suggest that the distribution of 'holes' follows a scaling relation between frequency and size, and that the boundaries of the clusters accord with space-filling lines.

Within this parameter space, it is clear that there are forms that accord with many types of urban settlement and, as such, this would appear a useful basis on which to begin some classification. What we propose is a thorough study of different growth situations in real cities with an emphasis on measuring speed of growth, average times of land development, and vacancy rates, thus linking our theoretical analysis to actual examples. We have not yet begun this, since the purpose of this chapter is simply to direct ideas in the study of growth and sprawl, and to propose a framework rich enough to capture the key elements of the development process. To begin our classification, Fig. 5.13 suggests at least a 2 x 2 set of types which define the ends of the spectrum across the two parameters: cities with low growth and low vacancy, probably characteristic of those which have grown slowly, maximizing their densities and with enough time to ensure that land is used efficiently – such cities might be found in some parts of Europe; cities which are even more compact, with high growth rates and low vacancy, might be characteristic of those in places like China; those with high vacancy rates and low growth rates, which lead to forms that are relatively compact but full of holes, would be like cities in western Europe which have undergone some form of de-industrialization; and cities with high growth rates and high vacancy rates which tend to sprawl like the cities of the American west.

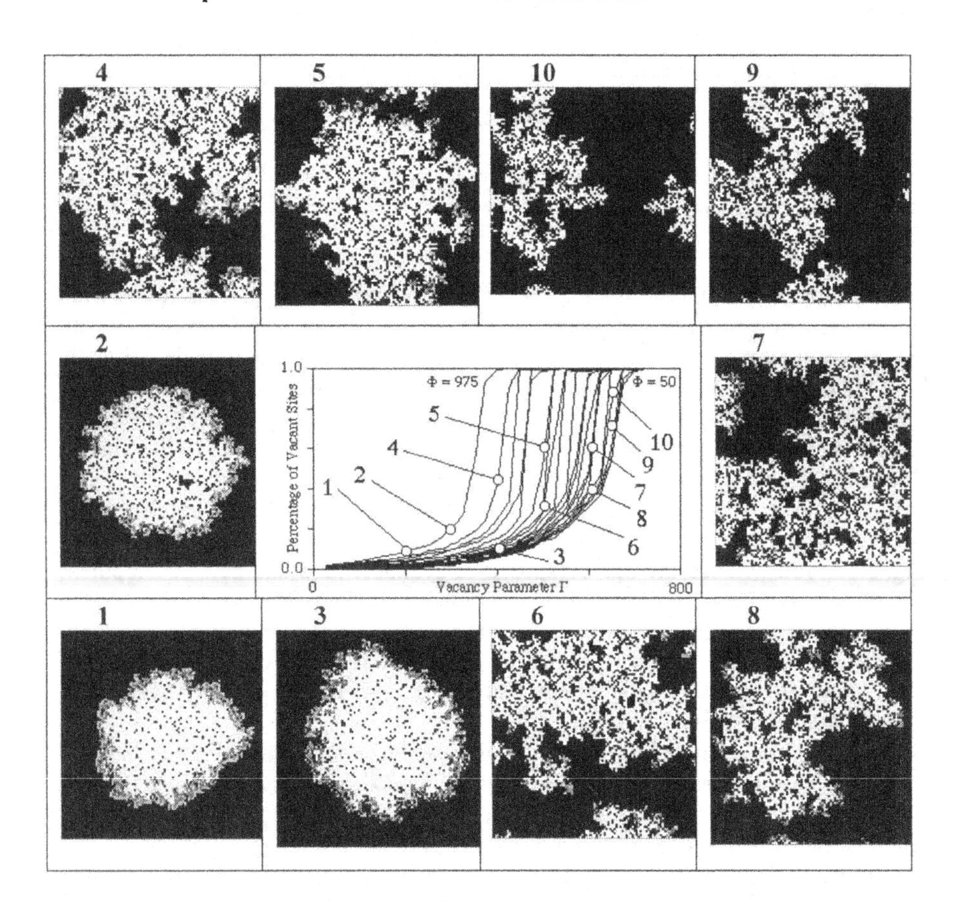

Fig. 5.13 Samples of growth from the parameter space $\{\Phi, \Gamma\}$

5.8 Conclusions: Applications and Policy

Although this chapter has been largely theoretical in tenor, we will conclude with some ideas about how these models might be focussed on more realistic applications. There are many variants of the framework that we have barely hinted at, in particular the generalization of these models to multiple seeds with the consequent merger of urban development as land around them passes through the growth process. Thus we will use the example of sprawl described here to show the effects of multiple seeds. We are currently developing an operational variant of this model for the town of Ann Arbor, MI, for which we have a rich database on how the region has developed over the last 20 years. To illustrate how the model might be applied, we take residential development from 1980 to 1985 and use this as the seeds for the development process. We then simulate this process from 1985 until 1990, for which we have observed data. In our empirical work, which will be reported in due course, our focus is on 'calibrating' the model over this time period and then making conditional predictions for the period 1990 to 1995. This will form an independent test of the model's ability to predict the dynamics of urban sprawl. Here however we will simply illustrate the model's ability to simulate the oscillations caused by the aging process of housing, with an emphasis on how 'initial conditions' – historical contingency – dictates the ultimate form of any and every urban system.

We report elsewhere a detailed CA model framework – **DUEM** [**D**ynamic **U**rban **E**volutionary **M**odelling] – which we are using for simulating sprawl (Batty, Xie and Sun 1999). This model has many more functions than the theoretical structure presented here, in that several land uses are simulated, neighbourhood sizes can be varied, and the transport network is predicted as a function of land use. In general terms every element can influence every other, but in the present demonstration we will restrict our focus exclusively to residential development, thus faithfully implementing the CA given in Equations (5.26) to (5.29) and (5.33) above. In Fig. 5.14(a), we show the urban extent of Ann Arbor in 1980 with the main highway structure, and in 14(b), residential development from 1980 until 1985, the seeds for our simulation.

The size of the grid on which this simulation takes place is 595 x 824 with each pixel about 30 square meters on the ground. We have used the 3 x 3 Moore neighbourhood with no constraints on existing development, with the result that ultimately the entire space is converted to new, then established development. Prior development is part of the development process which we assume is aged at $t = 0$ at the outset of the simulation.

The aging process assumed in **DUEM** involves each seed or site of new development being active or new for T_N time periods and then entering a mature phase in which it no longer determines new development for T_M time periods. After the period T_N+T_M, activity is redeveloped and re-enters the development process immediately. Currently, there is no vacancy within this model [although it can be simulated through a convolution of the aging process with variations in neighbourhood size] and thus the development that results is compact. In short, all land is developed in the long term except that reserved for the street network that acts a constraint. In the same way we illustrated for the CA models in Fig. 5.9 to Fig. 5.11, waves of new and established development, as well as redevelopment, move out concentrically around the 200 or so distinct seeds that form the initiators of the development process, as in Fig. 5.14(b). The 200 units of development which were created between 1980 and 1985, when used in the model process imply that each time unit in the model is equivalent to about 6 months in real time. In terms of the aging process, we assume that both T_N and T_M are set as 30 time units each. This implies that sites move through the development process to redevelopment in 30 years. This might be a little short for full-fledged redevelopment, although after this time there is a high probability of some form of renovation for all residential property.

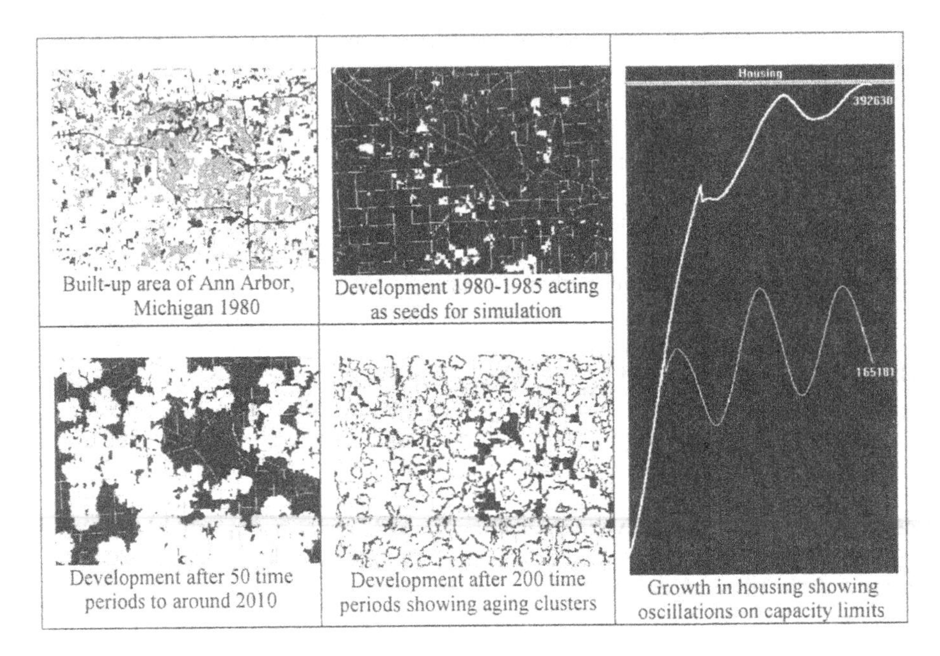

Fig. 5.14 Long term simulation of sprawl in the Ann Arbor region

Our simulation is thus one of very long term growth, two snapshots of which we show in Figs. 5.14(c) and 5.14(d). The first is growth after 50 time periods [about 25 years] which might reflect the situation in 2010 while the second is the simulation after 200 time periods [100 years] where the entire region has been developed and redeveloped several times. In both these scenarios, the narrow concentric bands of vacant sites show the paths of redevelopment around the initial seeds. The sole distinguishing morphological characteristic of the latter simulation is marked by these bands, thus reinforcing the notion that urban morphology is a combination of particular historical antecedents or accidents – the initial conditions – combined with the generalities of the development process. The oscillating behaviour of the development trajectories are shown in Fig. 5.14(e), where it is clear that the oscillations are muted until the system reaches its capacity and then the waves of redevelopment become quite distinct with respect to both new and established development. The phase and amplitude of these waves of course is a function of T_N and T_M.

In our quest to simulate the dynamics of the development process in the Ann Arbor region, we are introducing many more factors into our model. This will constrain its output to be more realistic as well as enable more realism to be simulated. For example, CA models do not usually contain constraints on the total activity generated, for it is assumed the attraction of sites and the probabilities of occupancy combine to determine the rate of growth. In real situations however, some limits on growth must be established. We achieve this by ensuring that the resulting distribution of development sites by size mirrors the actual scaling distribution, which we observe to be stable in time.

In developing these models, we are ultimately interested in policy applications. The set of models from which we have drawn inspiration in this chapter, has some quite well developed policy controls, the effects of which are well-known for epidemic-threshold models. For example, in countering the spatial diffusion of epidemics, particularly amongst animal populations, spatial controls on spread through traps of various sorts can be determined from the models. For example, in stopping the spread of rabies in foxes, Murray (1987) has shown how these models can be used to predict a belt of sufficient width that the transmission of the disease cannot cross it. Exactly the same logic can be developed in these urban models with respect to the imposition of green belts. In a way, this is already clear from the simulations made here, where redevelopment causes waves of vacancy to be established. In this sense, it is evident that if the development process is to be constrained through green belts, then 'artificial waves' might be introduced along the lines in which 'natural waves' occur within such models.

We have laid out the principles by which aging in the development process might be simulated and have moved theoretical models towards the kinds of practical applications which are urgently required for an improved understanding of the problems of urban sprawl. In future research, we shall take both the theoretical and practical consequences of such models further, particularly with respect to the detailed mechanisms of the development process and the kinds of urban morphology that these are able to reproduce.

Acknowledgements: This project was funded in part by the UK Office of Science and Technology Foresight Challenge [EPSRC GR/L54950]. Thanks are dude to Yichun Xie and Zhanli Sun of Eastern Michigan University who inspired the development of the DUEM software used in this chapter.

PART B: Spatial Application Domains

6 Spatial Pattern Recognition in Remote Sensing by Neural Networks

Graeme Wilkinson
School of Computing and Information Systems, Kingston University

6.1 Introduction

Neural networks are sophisticated pattern recognition tools. Over the last decade, following the early pioneering work of Benediktsson, Swain and Ersoy (1990) and Hepner et al. (1990), they have grown significantly in popularity as tools for the analysis of remote sensing data, primarily for the purpose of deriving spatial information used in environmental management. This popularity is due to several factors:

- Remote sensing applications are data rich [they rely increasingly on high dimensional imagery].
- Remote sensing data arise from complex physical and radiometric processes.
- There has been dissatisfaction with more conventional pattern recognition algorithms.
- There is a general desire to get maximum accuracy out of the data in remote sensing; high accuracy in some applications [e.g. agricultural subsidy fraud monitoring, mineral resource location] has high economic value.
- Neural networks are suitable for data for which statistical properties are unknown or poorly understood.
- Neural networks have potential scalability.
- Neural networks are potentially adaptable to parallel machine architectures.

Despite much optimism, and continuing exploration of neural network techniques in remote sensing, there has so far been no 'quantum leap' in the quality of products derived from remote sensing data through the adoption of neural network approaches. However, many reported results do demonstrate useful gains in accuracy in remote sensing products from use of neural networks in particular settings. At present, from the limited amount of evidence available, it is still too early to generalize about how and in what circumstances neural networks should be used in remote sensing. After a full decade of research, the fact that neural networks can be useful tools in remote sensing is, however, no longer in question as will be illustrated in the following sections.

6.2 Artificial and Biological Neural Networks

Although it is often claimed that artificial neural networks are modelled on biological neural systems, the fact remains that in reality, the functioning of biological neural systems is still extremely poorly understood and the ability of artificial neural systems to mimic them is highly restricted. Biological neural systems function through electrical and chemical processes taking place inside and between biological cells and membranes of enormous complexity. What is interesting, however, is that the biological systems continue to be a source of inspiration for the development of 'intelligent' mathematical algorithms, even though such systems are not completely understood.

In the case of neural networks, the overall 'architecture' of biological neural systems has been the inspiration for a variety of artificial neural network paradigms, though at a relatively superficial level. Whether this has any lasting technological value remains to be seen. Whilst the biological architectures may inspire mathematical models, it is still impossible to construct software programs or hardware devices that approach the complexity of real biological systems as currently understood. The human brain, for example, has something of the order of 10^{11} neurons and 10^{14} synaptic connections. Even the largest artificial neural networks only have of the order of 10^3 neurons with 10^4 interconnections – still a factor of 10 below that of the neural systems in the simplest invertebrates. Larger artificial systems run up against barriers of computability. Perhaps the most enduring legacy of research into artificial neural networks will be the fact that a whole class of algorithms are being developed. These are based around relatively simple processing elements which are connected in increasingly complex ways and between which information and signals are propagated. From such systems useful solutions to pattern recognition problems are being found in remote sensing, as in many other disciplines.

Taking a top-level view, it is clear that there are some useful analogies between what happens in satellite remote sensing and the behaviour of biological neural systems. For example, there are significant functional parallels between the human visual system and Earth observation. The human visual system is characterized by a sensory device [the eye] which receives a stimulus passed to the visual cortex and into the brain, where some form of understanding takes place. This understanding may result, for example, in the generation of a motor stimulus to enable the human to flee from a perceived threat. In the Earth observation context, the sensory device is normally a satellite spectro-radiometer, which gathers information about the global environment. This information is passed to data analysts who may use a variety of tools, including neural networks, in attempting to understand the meaning of the data. The response to this may, for example, be the formulation of public policies to control greenhouse gas emission. What is interesting about this comparison is that, whereas in the biological system the understanding and response may emerge relatively quickly [in seconds] from a single sensory-neural system [the brain], in the remote sensing context, the gathering and interpretation of the information may take many days or weeks [or even more]. Also, the actions resulting from that interpretation may take decades to implement through global political processes. Also, it is noteworthy that in the remote sensing context, the neural network pattern recognition algorithm is carrying out only a small part of the overall task of gathering information and acting on it.

6.3 Recent Developments in Remote Sensing

The technology of remote sensing is undergoing continual development. Changing technology, coupled with changing user requirements, has often been a driving force in the development and application of increasingly complex data analysis. In remote sensing, the recent technological changes have been (see, for example, Table 6.1):

- increasing numbers of spectral channels [up to 36 on recent sensors],
- improving spatial resolution [now 1m panchromatic, 4m multispectral],
- increased use of radar systems with multiple polarisation,
- increased use of mixed resolution data.

Table 6.1 Characteristics of some recent remote sensing instruments

Satellite / Sensor	Characteristics [Resolution]	Launch Date
SPOT-4: HRVIR /	4 bands vis-near IR [20m]	1998
VEGETATION	4 bands vis-near IR [1.15km]	
LANDSAT 7: ETM	6 bands vis-near IR [30m]	1998
	1 band thermal IR [60m]	
	Panchromatic [15m]	
TERRA: MODIS	21 bands vis-near IR [250m]	1999
	15 bands mid-far IR [1km]	
IKONOS-2	4 bands vis-near IR [4m]	1999
	Panchromatic [1m]	

Many remote sensing applications now rely on data gathered from more than one satellite. Data are also integrated together from multiple sensors, at multiple resolutions, and acquired on multiple dates [because of the additional discriminatory power offered by differences in the phenological cycles of different vegetation types]. In addition, such data may be combined with data sets from

Geographical Information Systems. Since each separate set of observations may be regarded as an independent mathematical feature, in combination they define a multidimensional data space in which analysis must take place. Whilst in the past, data set dimensionality may have typically been of the order of 4-6, current application problems exploiting the latest sensor systems, plus ancillary background GIS data, may require the analysis of data sets with dimensionality of the order of 50-100. This poses a significant problem for conventional multivariate statistical analysis, hence providing additional support for the exploration of alternative approaches such as neural networks.

6.4 Uses of Neural Networks in Remote Sensing

Neural networks first began to be used in remote sensing data analysis around 1990. Between 1990 and 1995 there was a rapid growth in applications due to increasing experimentation with architectures, paradigms and training methodologies. Besides their use in remote sensing, there has been gathering evidence of their value as general spatial analysis tools in solving more general problems in geography (Fischer and Gopal 1994a; Fischer 1998b).

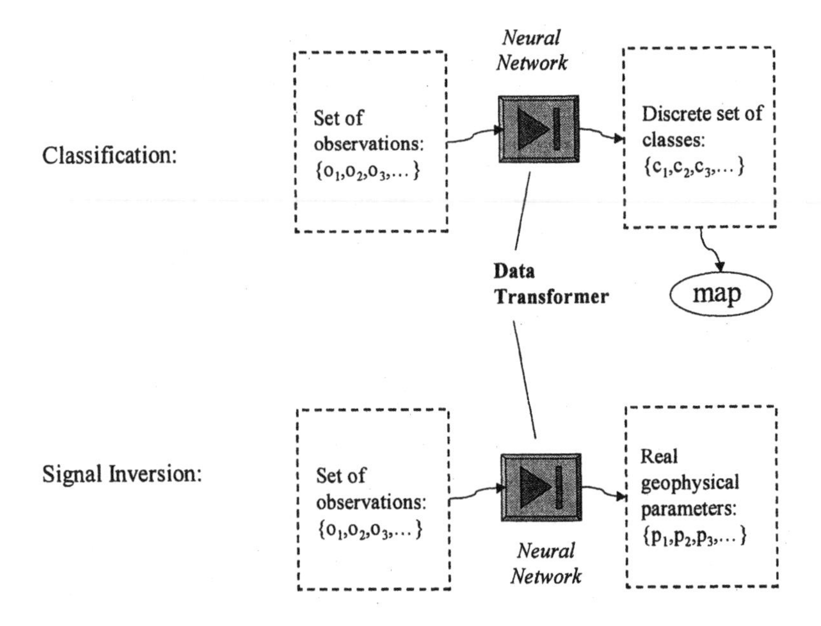

Fig. 6.1 Use of neural networks in remote sensing for classification and signal inversion. In both categories of application the neural network acts as a data transformer

Table 6.2 Reported neural network applications in [or closely related to] remote sensing

Type	Application	Example References
Classification	Land cover mapping	Chen et al. (95), Civco (1993), Dreyer (1993), Fischer et al. (1997); Kanellopoulos et al. (1992), Kavzoglu and Mather (1999), Paola and Schowengerdt (1997), Wilkinson et al. (1993), Yoshida and Omatu (1994), Schaale and Furrer (1995)
	Wetland mapping	Augusteijn and Warrender (1998)
	Urban mapping	Blamire and Mineter (1995)
	Rural area mapping	Bruzzone et al. (1997)
	Geological mapping	Yang et al. (1999), Gong (1996)
	Ship trail recognition	Clark and Boyce (1999)
	Tropical land use	Foody and Boyd (1999)
	Ecological and biodiversity mapping	Gong and Chen (1996)
	Cloud classification and masking	Lee et al. (1990), Logar et al. (1997), Walder and Maclaren (2000)
	Recognition of convective air masses	Pankiewicz (1997)
	Forest mapping	Skidmore et al. (1997), Wilkinson et al. (1995)
	Algal bloom mapping	Stephanidis et al. (1995)
	Lithologic mapping	Yang et al. (1998)
Signal Inversion	Phytoplankton concentration retrieval	Buckton, O'Mongain and Danaher (1999)
	Spectral mixture derivation	Foody et al. (1997)
	Biomass derivation	Jin and Liu (1997)
	Extraction of vegetation variables	Kimes et al. (1998)
	Forest backscatter retrieval	Kimes, Ranson and Sam (1997)
	Plant canopy backscatter inversion	Wang and Dong (1997), Gong, Wang and Liang (1999), Abuelgasim, Gopal and Strahler (1998), Pierce, Sarabandi and Ulaby (1994)
	Derivation of surface radiation fluxes	Schweiger and Key (1997)
	Cloud top height retrieval	Spina et al. (1994)
	Bulk-skin sea surface temperature difference retrieval	Ward and Redfern (1999)
	Rainfall estimation	Zhang and Scofield (1994)
Data Conditioning	Smoothing SAR data	Ellis, Warner and White (1994)
	Image segmentation	Haring, Viergever and Kok (1994)
	Data fusion	Wann and Thomopoulos (1997), Wan and Fraser (1999), Serpico and Roli (1995), Jimenez, Morales-Morell and Creus (1999)
	Image compression	Dony and Haykin (1995)
	Geometrical rectification	Di Carlo and Wilkinson (1996)

Although classification was the first major spatial application of neural networks in remote sensing, subsequent work demonstrated their value in signal inversion, i.e. the retrieval of geophysical parameters from complex sets of satellite observations (see Fig. 6.1). These two dominant types of application differ in that classification involves transforming satellite observations into categories, whereas signal inversion involves transforming observations into precise numerical parameters.

A third generic type of neural network usage in remote sensing has been 'data conditioning' – i.e. where the network is used in the pre-processing stage to improve the quality of the data, or to rectify it in some way [e.g. geometrically]. Examples of different applications within the overall broad groups of 'classification', 'signal inversion', and 'data conditioning' are given in Table 6.2.

Regardless of which kind of generic usage is considered, remotely sensed data are inherently spatial data, and their transformation into meaningful products is a spatial problem. Furthermore, in most cases, the spatial structure of the derived products is an important aspect and the neural network approach must provide the spatial patterns required by the end user.

6.5 Creation of Neural Network Input Vectors

In the vast majority of applications of remote sensing, analysis is performed with a multidimensional image data set supplemented by ancillary geographically co-registered GIS data sets. Analysis is therefore based on the multiple items of information available from each of a number of data layers concerning a point on the surface of the Earth. Input feature vectors for neural network input are thus created by streaming data from the data layers, creating a so-called 'stack vector', see Fig. 6.2.

In some cases, the stack vector will be augmented by observations from the vicinity of the ground point of interest, e.g. where the local texture of an image region may be useful in classifying the land use type. In other cases, parameters may be derived from the various data layers and used to augment the stack vector in some way. A typical example of this would be in calculating the local Normalised Difference Vegetation Index [NDVI] and using it as an additional feature in classification.

6.6 Neural Networks in Unsupervised Classification of Remote Sensing Data

Classification, for the purpose of spatial thematic mapping, has continued to be one of the main uses of neural networks in remote sensing. Unsupervised classification involves the process of identifying the main clusters in complex remote sensing data sets, labelling them, and training a classifier algorithm to recognize further examples of the same clusters drawn from new 'unseen' data (see Fig. 6.3).

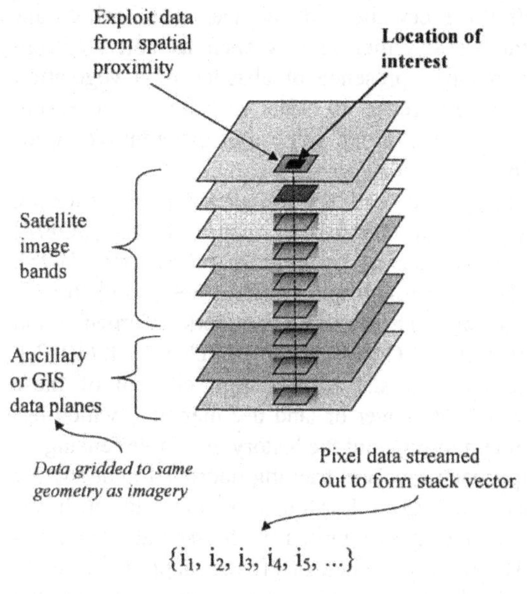

Fig. 6.2 Creation of neural network input 'stack vector' from remote sensing and GIS data layers

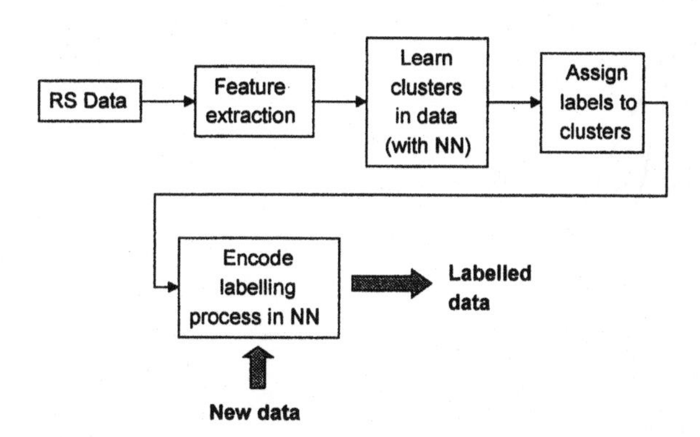

Fig. 6.3 Unsupervised classification procedure in remote sensing

One of the main reasons why unsupervised classification is so popular in remote sensing is that raw satellite radiometric measurements are rarely fully understood. Differences in terrain and vegetation canopy conditions can significantly affect satellite observation. Two seemingly identical apple orchards may appear very different in satellite imagery due to differences in the soil colour or texture below and between the trees. Other factors such as soil moisture content, slope and aspect of the orchards, presence or absence of a vegetation understory, can make a significant difference to what is actually observed. Clustering procedures may therefore help to group somewhat different examples of the same basic land use type into the same category.

Several types of neural network can be used for unsupervised classification (Schalkoff, 1997). These include the Self-Organizing Feature Maps [SOFM] (Kohonen, 1988) [as applied in remote sensing by Schaale and Furrer (1995), Wilkinson, Kontoes and Murray (1993), Yoshida and Omatu (1994), and Ito and Omatu (1997)] and the Adaptive Resonance Theory [ART] networks (Carpenter and Grossberg 1987b) [as applied by Fischer and Gopal (1993b), Fischer et al. (1997)]. With the exception of Wilkinson, Kontoes and Murray (op. cit.) all of these applications have been concerned with land cover or land use mapping, which has remained a notoriously difficult problem throughout the history of remote sensing.

In essence, the unsupervised approach involves training neurons to encode the characteristics of the principal clusters in a given Earth observation data set. In the Kohonen approach, a multidimensional array of neurons [with associated weights] is regarded as a 'mapping cortex'. By training the neurons in the mapping cortex to resemble the main data clusters, the 'codebook' of vectors associated with the neurons comes to be an efficient representation of the variability in the raw data set (see Fig. 6.4).

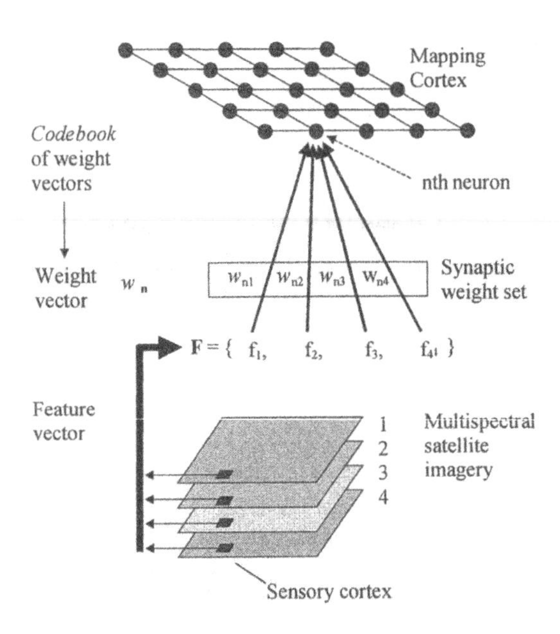

Fig. 6.4 Kohonen self-organizing feature map

The training procedure for SOFM networks is illustrated in Fig. 6.5. The array of neurons carry an initially random set of weights [as weight vectors]. As each sample from the remote sensing data set is presented to the net, the closest 'winner' neuron is identified and its weight vector [and all others within a defined neighbourhood] are adjusted to be more like the sample's feature vector. [Note that the weight vector must have the same dimensionality as the remote sensing data feature vector].

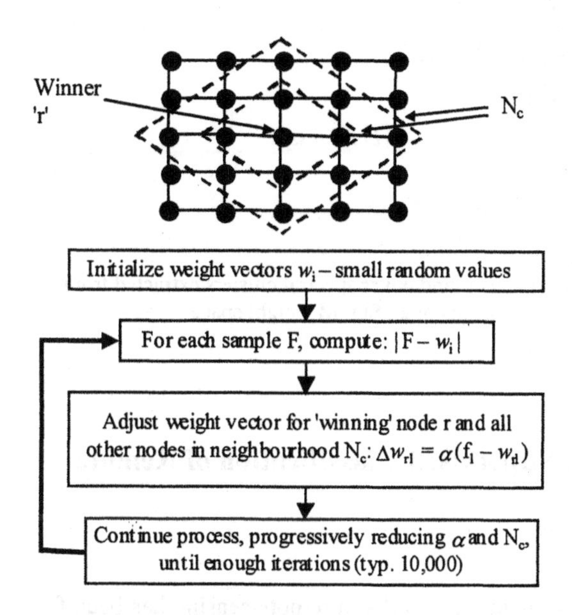

Fig. 6.5 Training procedure for Kohonen SOFM

The neurons in the mapping cortex of a SOFM, progressively migrate in weight space during training. Their final positions provide a mathematical map of the various data clusters in the data set and their spectral similarity (Fig. 6.6). The SOFM approach has been highly successful in remote sensing applications to date. Interestingly, Wan and Fraser (1999) have extended the basic SOFM approach into a more comprehensive data fusion, clustering, and classification architecture for remote sensing. Their approach uses joint multiple SOFMs [*j MSOFMs*] which operate on satellite data from different dates. Also each SOFM uses an *augmented* input vector [*a MSOFM*], which extends the basic spectral observations with contextual co-occurrence vectors [i.e. providing information about the spatial context of pixel samples]. Moreover, and of even greater interest, their method uses the provisional cluster labels of pixel samples as an additional input feature on successive training iterations in a recursive process. Their so-called '*a MSOFM/j MSOFM*' approach has considerable flexibility and appears to offer much potential in remote sensing, although further experimentation with such architectures is required.

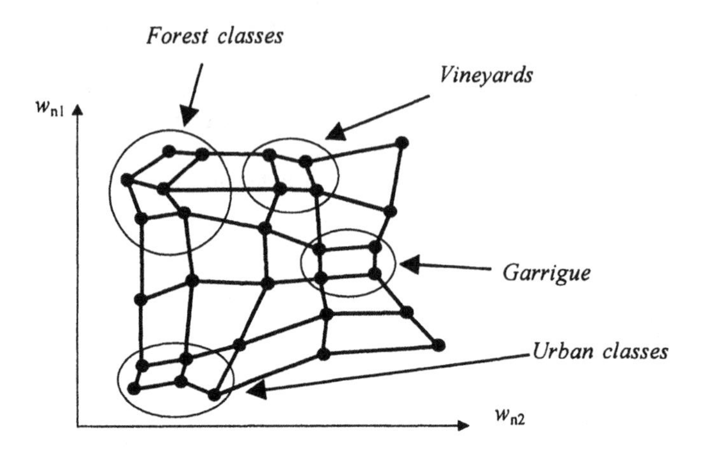

Fig. 6.6 Simulated distribution of groups of land cover spectral clusters extracted from satellite imagery as appearing in two dimensions of an SOFM weight space

6.7 Neural Networks in Supervised Classification of Remote Sensing Data

By far the most common usage of neural networks in remote sensing has been for supervised classification, in which known ground truth information is used to train a neural classifier to recognize classes of interest in satellite data. This approach has been applied extensively to land surface classification for many spatial purposes, such as mapping of general land cover, of more specific urban zones, forest areas, rural areas and agricultural fields. The main objectives of any supervised classification algorithm are to be able to learn the characteristics of training data efficiently, and also to be able to accurately generalize from its learning in order to classify unseen data without error. The process relies on the availability of good training data – usually created by ground surveys and site inspections to identify example pixels in the satellite images of required terrestrial mapping classes (Fig. 6.7).

The most commonly applied neural network approach for supervised classification in remote sensing has been the multilayer perceptron (Fig. 6.8) trained with the backpropagation algorithm. This approach has been comprehensively reviewed by Paola and Schowengerdt (1995). Multilayer perceptrons act as data transformers and learn to encode a division of feature space into separate user-defined classes.

Although it has now been used quite extensively and has often been found to yield more accurate results than conventional statistical classifiers (such as the maximum-likelihood method, e.g. Downey et al. 1992), there remain many aspects of this approach which require deeper understanding and investigation. Some pointers will be given below to a number of the critical issues in remote sensing which currently need further development. An important and noteworthy feature of multilayer perceptron networks [MLPs] is that they are *semi-linear* – an aspect which has certain consequences in remote sensing. The semi-linear aspect of MLPs can be understood by examining the behaviour of individual neurons (Fig. 6.9).

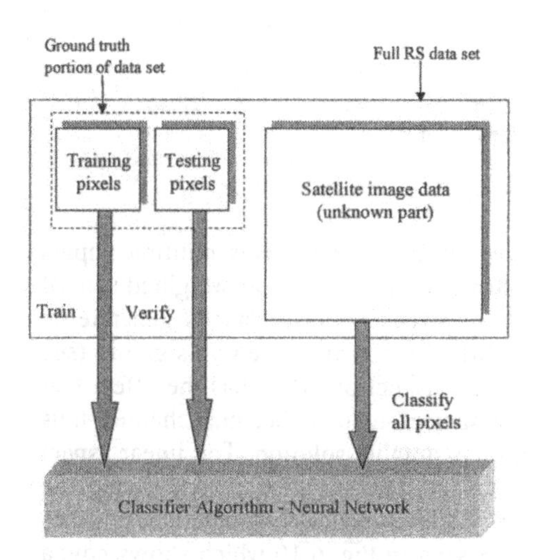

Fig. 6.7 Training and verification data sets for classification

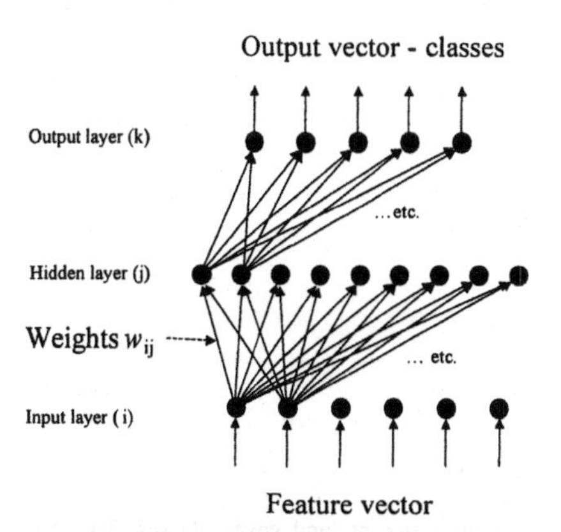

Note: In some applications the feature vector may have significantly more componentes than illustrated. Likewise, some applications will require the derivation of many more classes.

Fig. 6.8 Structure of a multilayer perceptron network

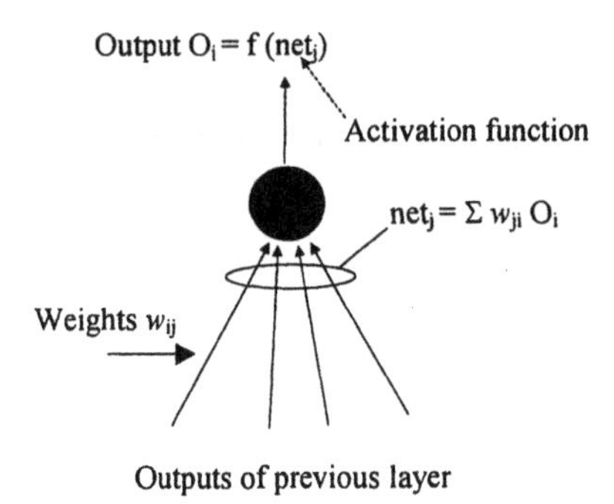

Fig. 6.9 Single neuron in a multilayer perceptron network

Each neuron, in all but the input layer, in an MLP receives multiple inputs from the preceding layer. The neuron then initially forms a linear weighted sum of these before applying an activation function to the result in order to generate its output signal. The activation function is normally non-linear, e.g. a sigmoid (see for example, Schalkoff, 1997). The non-linear aspect of MLPs has the effect that the training process can in certain circumstances tend to become chaotic, thus preventing the network from converging on a suitable solution. The linear aspect of the MLP has the effect of making the decision surfaces, which divide the classes, take on linear [or more accurately hyper-planar] characteristics in the multidimensional feature space. This can be seen in Fig. 6.10 which shows how a two-dimensional remote sensing feature space is divided into land cover classes by a trained MLP.

Note: The different shades represent zones corresponding to different land cover classes. The two dimensional feature space corresponds to the pixel values in bands 4 and 5 of the Landsat 5 Thematic Mapper instrument. This 2D feature space is actually a slice through a six dimensional space in which learning and classification occurs (after Fierens et al. 1994).

Fig. 6.10 Division of two-dimensional feature space by a MLP

One of the most difficult problems in the classification of remotely sensed data is that the process usually takes place in a space off ar more than two dimensions, due to the number of features extracted from the data set and included in the 'stack vector'. This number is also tending to grow, as indicated earlier, due to the increasing sophistication of the sensor systems now being developed and flown in space. It is therefore extremely difficult to visualise the data space and to examine geometrically how different forms of classifier behave. Some attempts have been made to aid this process by using virtual reality for immersive three-dimensional visualisation (Petrakos, Di Carlo and Kanellopoulos 1999; Di Carlo, 2000). However, even this helps only to a small degree, as there is still the need to reduce data set dimensionality from, in most cases, many more than three dimensions by slicing or projecting the data. Visual understanding of the mathematical behaviour of the classifier in the full feature space is thus severely constrained and behaviour in two or three dimensions can only be seen as an indication of what might be happening in other dimensions. For this reason, neural networks tend to be regarded in remote sensing as 'black boxes' into which training data are fed and from which a division of feature space emerges. Success is judged by whether the division of feature space achieves a required level of accuracy when unseen verification data are introduced to the net. Likewise, inter-comparisons of different neural network models and architectures also rely on the same testing approach – i.e. on the introduction of previously unseen data and the evaluation of the classification results (e.g. in terms of the percentage of pixels from a given ground truth class being correctly assigned to that same ground truth class or by analysis of confusion matrices, Congalton 1991).

Regrettably, since there are as yet no standard data sets for evaluating classifier performance in remote sensing, reported experiments have all tended to use their own highly specific data sets, and hence the results of different experiments performed by different workers usually cannot be realistically compared.

6.8 'Soft Computing' Approaches Using Neural Networks

The traditional remote sensing classification approach involves assigning pixels in satellite images into one of a number of fixed discrete classes – an approach for which MLPs, and other neural networks, have been employed. However, it is becoming increasingly recognized that it is extremely difficult in many cases to give unique, unambiguous, discrete class labels to areas of the land surface. When presented within previously unseen satellite data, MLP output layer neurons do not neatly carry a '1' on the node representing the preferred class label and '0' on all of the others. The output layer suggests a mixture of classes with varying strengths (Fig. 6.11). This has given rise to 'soft classification' approaches in which fuzzy or mixed classification is used for regions of satellite data sets (Fig. 6.12).

Neural networks can be used to perform soft classification in several ways. One is simply to interpret the normalized relative strengths of the signals from the output neurons as indicators of the fuzzy proportions of the fixed discrete classes which the network has been trained to recognize. Another is to train the classifier with examples of mixed classes and to adopt the 'winner take all' approach, i.e. assign the data sample to the single mixed class which has the highest strength at the network output layer. A more sophisticated approach would be to derive *fuzzy proportions* of *mixed classes*. Note that any of these approaches are acceptable. The real value of the results can only be determined in the context of the use of the remote sensing data – are the results, whether hard, fuzzy, mixed, or fuzzy-mixed, useful? Clearly there are network training implications arising from the type of classification employed. Mixed classification requires training with examples of the mixed classes. Training for fuzzy classification on the other hand can be performed only with examples of hard classes as hard and fuzzy classifiers can be the same internally [the only difference being in the way in which the output layer signals are interpreted]. However, training for fuzzy classification can also be performed with fuzzy training data [i.e. where a training sample is regarded as being a fuzzy class sample and the target output vector for that sample has normalised real values representing the fuzzy membership of each class].

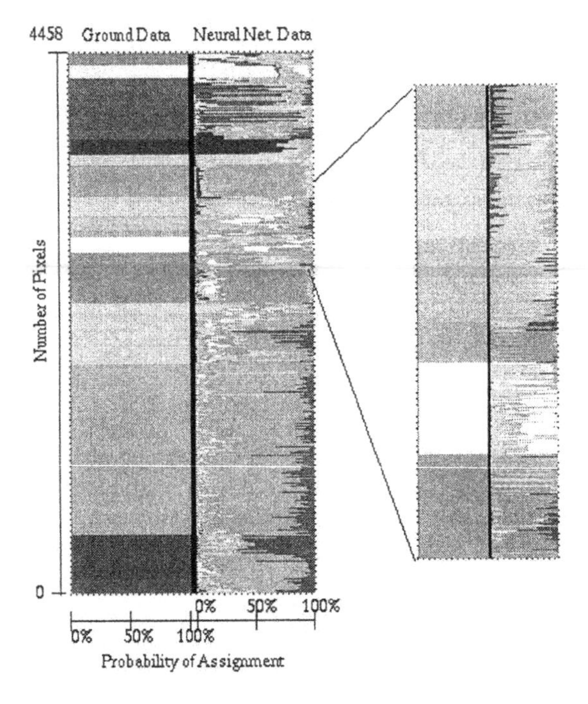

Fig. 6.11 Complexity of neural network output state compared to single hard classification of ground data (after A. Bernard, EU Joint Research Centre, Ispra)

Interestingly, work by Bernard, Wilkinson and Kanellopoulos (1997) has shown that it can be beneficial to train an MLP with fuzzy training data. This appears to improve the accuracy, even of hard classification. There are many examples of fuzzy or soft classification using neural networks in the remote sensing literature, e.g. Gopal, Sklarew and Lambin (1993), Zhang and Hoshi (1994), Foody (1996), and Warner and Shank (1997). A recent extension of the soft computing approach has been to use a fuzzy system to combine the results of multiple network classifications when dealing with complex problems (Thackrah, Barnsley and Pan 1999).

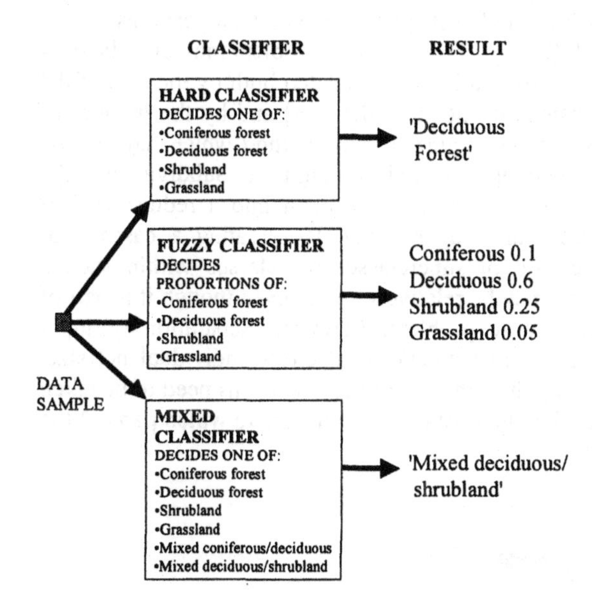

Fig. 6.12 Different approaches to supervised classification in remote sensing: Hard, fuzzy and mixed (note that fuzzy classification can also be applied to mixed classes yielding 'abundances' or 'mixture proportions' *of mixed classes*)

6.9 Managing Complexity

As indicated earlier, one of the current trends in remote sensing is to use increasingly sophisticated sensor systems – with higher resolutions and increasing numbers of spectral channels [e.g. 'hyperspectral' systems]. At the same time, users require ever more precise environmental products covering wider and wider areas of terrain [in some cases covering whole continents]. In some applications, to achieve viable products, it is now desirable to use data from different dates [especially for vegetation] and captured by different sensors at multiple viewing angles. Overall, the net effect of such trends is to vastly increase the complexity of the classification or signal inversion task. This ultimately requires larger and more complex neural networks, which can become too large to train in a realistic time-scale. So far there have been relatively few studies devoted to neural network classification of data sets with high dimensionality (for some examples see: Benediktsson, Swain and Ersoy 1993; Benediktsson and Kanellopoulos, 1999; and Jimenez, Morales-Morell and Creus 1999). Essentially there are two ways in which increasing complexity can be managed in remote sensing data analysis with neural networks. One is to subdivide the problem and to use multiple networks [or other classifiers] in tandem, working on separate parts. The other is to devise ways of speeding up the training of large single nets by use of parallel algorithms.

In the sub-division approach (Fig. 6.13), a complex problem may be solved by a set of relatively small networks. Network size is dictated both by the size of the input feature vector [i.e. dimensionality of input data set] and by the number of output neurons [i.e. number of classes required]. Any method which reduces the size of these can reduce overall network size and training time. Subdivision at the output end requires separating the classification problem into a reduced set of classes or a hierarchy of classes and superclasses. By creating a cascade of networks to handle a restricted set of superclasses [or classes within such a superclass] an individual network may be relatively small, even though it is part of a combined solution to a very large problem. Likewise, separation of input features and feeding them to different networks can reduce individual net size, though this approach is more difficult in that intelligent decisions need to be taken about which parts or features of the input data set are relevant to which parts of the overall problem.

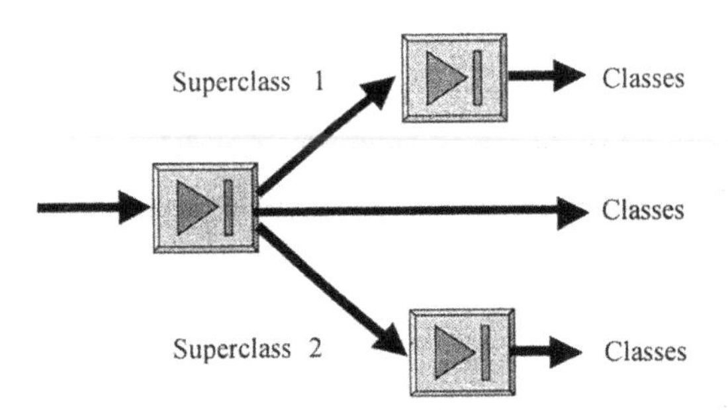

Fig. 6.13 Problem subdivision in neural network classification

The second approach to complexity relies on the use of parallel processing. Even though inherent parallelism has been one of the driving forces behind the adoption of neural network approaches in many disciplines, the number of applications in which parallelism has actually been used, either in software or hardware, has been relatively small.

The rationale for using parallel processing is very clear. The training of neural networks can be computationally prohibitive. For an MLP network, with N_i, N_h, and N_o nodes in its input, single hidden, and output, layers respectively, the number of floating point operations for one forward and one backward pass in training is approximately $2N_h(2N_i+N_o)$ multiplications/additions plus (N_h+N_o) sigmoids where 1 sigmoid is computationally equivalent to approximately 14 multiplications (Saratchandran, Sundararajan and Foo 1996). For a network of 50 neurons and a training set of 1000 samples, there would be ~15 million floating point operations per epoch in batch training – without weight update! There are two approaches to parallelising network training as shown in Fig. 6.14. These rely on either splitting the training data between different networks or dividing the overall network between different computational processors.

Besides adopting parallelism, it is likely that recent developments in methodologies to improve the effectiveness with which network architectures and weight sets can be algorithmically determined will also help. For example, the proposed use of genetic algorithms to evolve optimal architectures and weight sets for modelling spatial data (Fischer and Leung 1998) is particularly interesting.

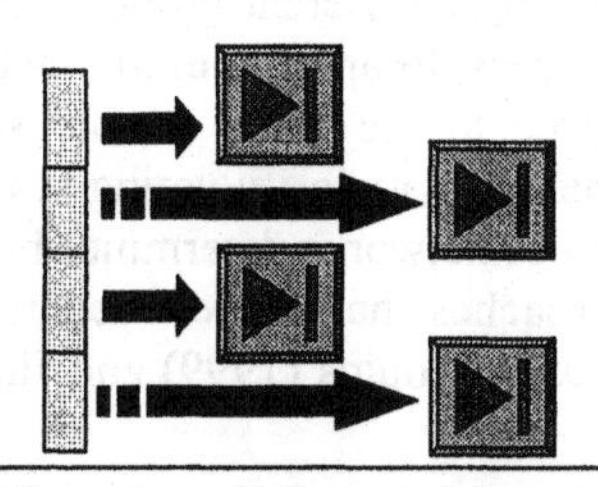

Training set parallelism - training set split between identical copies of same net run on different processors. Weight updates shared at end of each epoch.

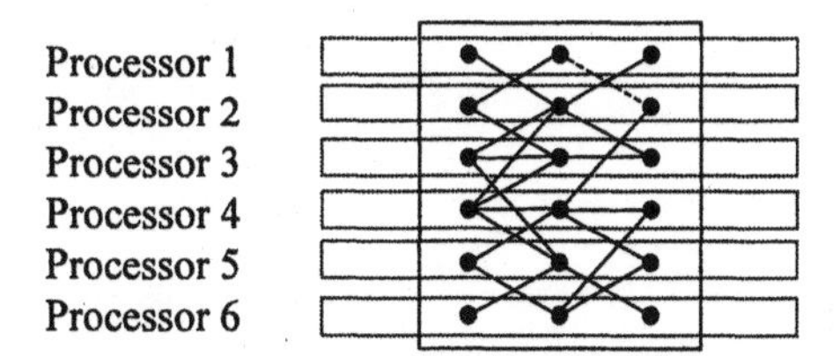

Network parallelism - processors handle different parts of the network

Fig. 6.14 Two approaches to neural network parallelism speed up training of large complex networks

6.10 Hybrid Analysis Methodologies

In the previous section, we explored the possibility of subdividing a problem into separate networks for efficiency purposes. However, the multiple network approach has also been found useful as a means of improving overall accuracy. Neural networks actually only approximate the desired solution to any data analysis problem. Network training involves starting with randomized weight sets and progressively adjusting weights to yield an acceptable division of the input feature space. Networks progressively learn by following a strategy aimed at minimizing their overall error e.g. using gradient descent or conjugate gradient methods in backpropagation (see for example, Schalkoff 1997).

However, different networks starting from different initial points in weight space can finish their training with quite different divisions of the feature space. Some may be more accurate in recognizing certain classes in a remote sensing data set, others may be more accurate in recognizing a different group of classes. Several attempts have been made, with some success, to apply multiple networks encoding different divisions of the feature space to the same remote sensing problem. The key to interpretation then lies in having a majority voting system or a consensus system in which the collective group of networks determines between them a unique preferred answer. Such approaches have been reported by Benediktsson et al. (1997), Benediktsson and Kanellopoulos (1999) and Jimenez, Morales-Morell and Creus (1999).

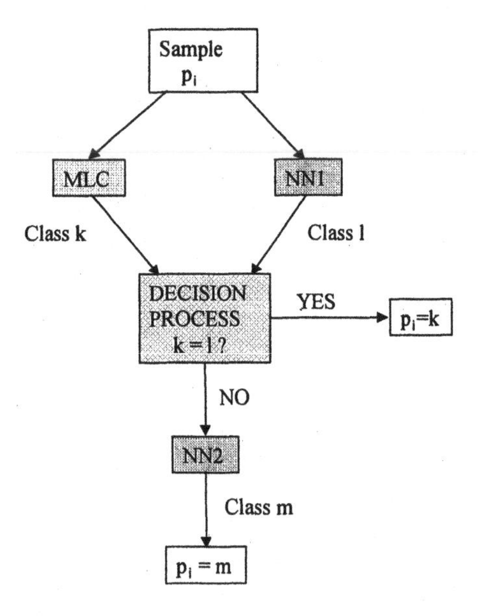

Fig. 6.15 Simple hybrid statistical neural classifier (after Wilkinson et al. 1995)

Finally, it is worth noting that in some cases it may be preferable to combine neural networks with other techniques. Statistical classifiers, such as the maximum-likelihood approach, divide the feature space with intersecting multidimensional ellipsoidal equi-probability surfaces. These are very different to the quasi-hyper-planar surfaces empirically found with neural networks such as MLPs. Wilkinson et al. (1995) were able to improve overall classification accuracy by jointly using a maximum likelihood classifier [MLC] and two neural nets (Fig. 6.15).

Initially the MLC and Neural Net 1 both classified a sample. If they agreed, the result was settled. If they disagreed, the sample was passed to Neural Net 2 which made the classification decision. Neural Net 2 was explicitly trained with pixels for which the MLC and Neural Net 1 did not agree and thus became attuned to 'difficult pixels' – which were possible outliers from the main cluster centres that neither the MLC nor Neural Net 1 could adequately deal with.

Besides integrating statistical and neural network approaches, there has also been interest in combining multiple pattern analysis paradigms to include not only statistical and neural systems, but also fuzzy systems and expert systems. The integration of expert systems with neural networks is still a very new area with few results to date, although the potential of such integration in spatial analysis within geographical information systems has already been identified by Fischer (1994). It may be appropriate in future to solve extremely complex computational problems in remote sensing by a combination of such techniques. Expert systems have the advantage of being constructed from explicitly stated logical rules, which can encode human knowledge and be understood by humans. By contrast neural networks are 'opaque' mathematical systems whose behaviour in a multidimensional mathematical space can be very difficult to understand. The most likely outcome in the future is that highly complex spatial problems in remote sensing will be solved with hybrid systems, which exploit expert systems for some aspects of the overall problem and neural networks for others (Fig. 6.16). This still remains a formidable scientific challenge.

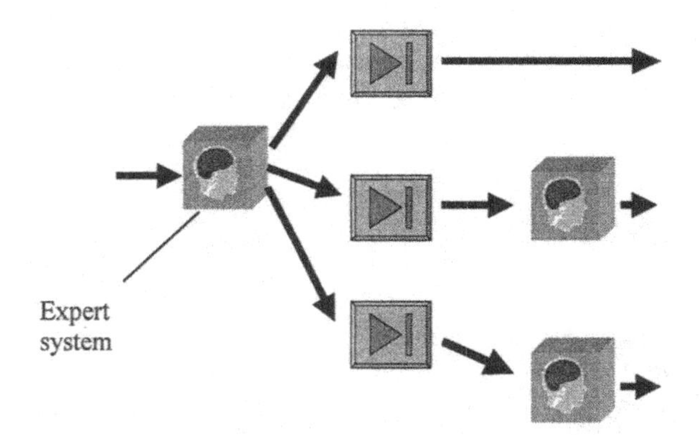

Fig. 6.16 A potential combination of expert systems and neural networks for solving a complex pattern recognition problem

6.11 Conclusions

Neural networks are now well established as an analytical tool for spatial pattern recognition in remote sensing. They have found application in signal inversion and in classification – both supervised and unsupervised. It would thus seem that they have a key role to play in the future development of the science of remote sensing. As remotely sensed data sets become more complex, however, more research will be needed to use them efficiently and effectively. Training times are becoming more of an issue with the growing use of hyperspectral data sets and accuracy is becoming more critical as remote sensing is increasingly driven by commercial applications for which accuracy has economic value. For this reason, the dominant approach in the future is likely to be the use of hybrid and parallel systems in which complex remote sensing data analysis problems will be decomposed into solvable or computable sub-problems which are acted on by different sub-systems – multiple neural networks, expert systems, and fuzzy systems. Ultimately such systems may be viewed as emergent intelligent systems, which perform the task of transforming an enormously complex data space into meaningful maps or spatial patterns of required geophysical parameters.

7 Fuzzy ARTMAP – A Neural Classifier for Multispectral Image Classification

Sucharita Gopal and Manfred M. Fischer***
* Department of Geography, Boston University
** Department of Economic Geography & Geoinformatics,
Vienna University of Economics and Business Administration

7.1 Introduction

Spectral pattern recognition deals with classifications that utilize pixel-by-pixel spectral information from satellite imagery. The literature on neural network applications in this area is relatively new, dating back only about six to seven years. The first studies established the feasibility of error-based learning systems such as backpropagation (see Key, Maslanik and Schweiger 1989; McClellan et al. 1989; Benediktsson, Swain and Ersoy 1990; Hepner et al. 1990). Subsequent studies analyzed backpropagation networks in more detail and compared them to standard statistical classifiers such as the Gaussian maximum likelihood (see Bischof; Schneider and Pinz 1992; Kanellopoulos, Wilkinson and Mégier 1993; Fischer et al. 1994).

In this chapter[1] we analyze the capability and applicability of a different class of neural networks, called fuzzy ARTMAP, to multispectral image classification. Fuzzy ARTMAP synthesizes fuzzy logic and Adaptive Resonance Theory [ART] models by describing the dynamics of ART category choice, search and learning in terms of analog fuzzy set-theoretic rather than binary set-theoretic operations. We describe the design features, system dynamics and simulation algorithms of this learning system, which is trained and tested for classification of a multispectral image of a Landsat-5 Thematic Mapper [TM] scene [270x360 pixels] from the City of Vienna on a pixel-by-pixel basis. Fuzzy ARTMAP performance is compared with that of a backpropagation system and the Gaussian maximum likelihood classifier on the same database.

The chapter is organized as follows. Section 7.2 gives a brief mathematical description of the unsupervised learning system, called ART 1, which is a prerequisite to understanding the ARTMAP system. Section 7.3 shows how two ART 1 modules are linked together to form the ARTMAP supervised learning system for binary pattern recognition problems. Section 7.4 introduces a generalization of ARTMAP, called fuzzy ARTMAP, that learns to classify continuous valued rather than binary patterns, and presents a simplified version of the general fuzzy ARTMAP learning system, which will be used as general purpose remote sensing classifier in this study. Section 7.5 describes the remote sensing classification problem which is used to test the classifier's capabilities. The simulation results are given in Section 7.6 and compared with those obtained by the backpropagation network and the conventional maximum likelihood classifier. In the final section, we draw some conclusions.

7.2 Adaptive Resonance Theory and ART 1

The basic principles of adaptive resonance theory [ART] were introduced by Stephen Grossberg in 1976 as a theory of human cognitive information processing (Grossberg 1976a, b). Since that time, cognitive theory has led to a series of ART neural network models for category learning and pattern recognition. Such models may be characterized by a system of ordinary differential equations (Carpenter and Grossberg 1985, 1987a) and have been implemented in practice using analytical solutions or approximations to these differential equations.

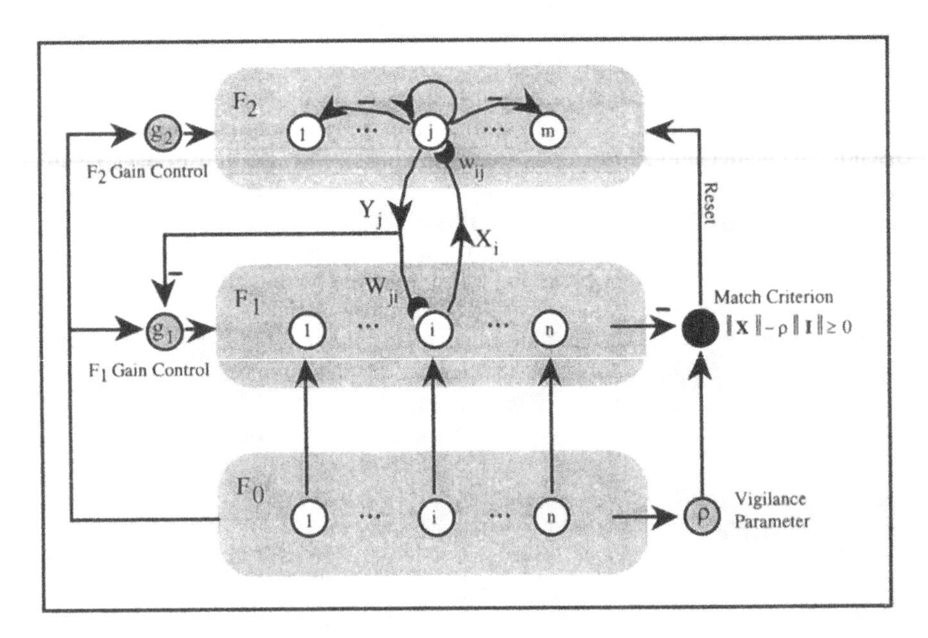

Fig. 7.1 The ART 1 architecture

ART models come in several varieties, most of which are unsupervised, and the simplest are ART 1 designed for binary input patterns (Carpenter and Grossberg 1987a) and ART 2 for continuous valued [or binary] inputs (Carpenter and Grossberg 1987b). This section describes the ART 1 model which is a prerequisite to understanding the learning system fuzzy ARTMAP. The main components of an ART 1 system are shown in Fig. 7.1. Ovals represent fields [layers] of nodes, semicircles adaptive filter pathways and arrows paths which are not adaptive. Circles denote nodes [processors], shadowed nodes the vigilance parameter, the match criterion and gain control nuclei that sum input signals. The F_1 nodes are indexed by i and the F_2 nodes by j [categories, prototypes]. The binary vector $\mathbf{I}=(I_1, ..., I_n)$ forms the bottom-up input [input layer F_0] to the field [layer] F_1 of n nodes whose activity vector is denoted by $\mathbf{X}=(X_1, ..., X_n)$. Each of the n nodes in field [layer] F_2 represents a class or category of inputs around a prototype [cluster seed, recognition category] generated during self-organizing activity of ART 1. Adaptive pathways lead from each F_1 node to all F_2 nodes [bottom up adaptive filter], and from each F_2 node to all F_1 nodes [top down adaptive filter]. All paths are excitatory unless marked with a minus sign.

Carpenter and Grossberg designed the ART 1 network using previously developed building blocks based on biologically reasonable assumptions. The selection of a winner F_2 node, the top down and bottom up weight changes, and the enable/disable [reset] mechanism can all be described by realizable circuits governed by differential equations. The description of the ART 1 simulation algorithm below is adapted from Carpenter, Grossberg and Reynolds (1991a) and Carpenter, Grossberg and Rosen (1991b). We consider the case where the competitive layer F_2 makes a choice and where the ART system is operating in a fast learning mode.

F_1-Activation

Each F_1 node can receive input from three sources: the $F_0 \rightarrow F_1$ bottom-up input; non-specific gain control signals; and top-down signals from the m nodes [winner-take-all units] of F_2, via an $F_2 \rightarrow F_1$ adaptive filter. A node is said to be *active* if it generates an output signal equal to 1. Output from inactive nodes equals 0. In ART 1 a F_1 node is active if at least two of the three input signals are large. This **rule for F_1 activation** is called the **2/3 Rule** and realized in its simplest form as follows: The ith F_1 node is active if its net input exceeds a fixed threshold:

$$X_i = \begin{cases} 1 & \text{if } I_i + g_1 + \Sigma_{j=1}^{n} Y_j W_{ji} > 1 + \overline{W} \\ 0 & \text{otherwise,} \end{cases} \tag{7.1}$$

where I_i is the binary $F_0 \rightarrow F_1$ input, g_1 the binary non-specific F_1 gain control signal, and term $\Sigma_{j=1}^{n} Y_j W_{ji}$ the sum of $F_2 \rightarrow F_1$ signals Y_j via pathways with adaptive weights W_{ji}, and $\overline{W}$ $(0<\overline{W}<1)$ is a constant. Hereby the F_1 *gain control* g_1 is defined as

$$g_1 = \begin{cases} 1 & \text{if } F_0 \text{ is active and } F_2 \text{ is inactive} \\ 0 & \text{otherwise.} \end{cases} \tag{7.2}$$

It is important to note that F_2 activity inhibits g_1, as shown in Fig. 7.1. These laws for F_1 activation imply that, if F_2 is inactive, then

$$X_i = \begin{cases} 1 & \text{if } I_i = 1 \\ 0 & \text{otherwise.} \end{cases} \tag{7.3}$$

If exactly one F_2 node J is active, the sum $\Sigma_{i=1}^{n} X_i \ W_{ji}$ in Equation (7.1) reduces to the single term W_{Ji}, so that

$$X_i = \begin{cases} 1 & \text{if } I_i = 1 \text{ and } W_{Ji} > \overline{W} \\ 0 & \text{otherwise.} \end{cases} \tag{7.4}$$

Rules for Category Choice [F_2 choice]

F_2 nodes interact with each other by lateral inhibition. The result is a competitive winner-take all response. The set of committed F_2 nodes [prototypes] is defined as follows. Let T_j denote the total input from F_1 to the jth F_2 processor, given by

$$T_j = \Sigma_{i=1}^{n} \ X_i \ w_{ij} , \tag{7.5}$$

where w_{ij} represent the $F_1 \rightarrow F_2$ [i.e. bottom-up or forward] adaptive weights. If some $T_j > 0$, define the F_2 choice index J by

$$T_J = \max_{j=1,\ldots,m} \{T_j\}. \tag{7.6}$$

Characteristically, J is uniquely defined. Then the components of the F_2 output vector $\mathbf{Y} = (Y_1, \ldots, Y_m)$ are

$$Y_i = \begin{cases} 1 & \text{if } j = J \\ 0 & \text{if } j \neq J. \end{cases} \tag{7.7}$$

If two or more indices j share maximal input, then one of these is chosen at random.

Learning Laws: Top Down and Bottom Up Learning

The *learning laws* as well as the rules for choice and search, may be described, using the following notation. Let $\mathbf{A} = (A_1, ..., A_m)$ be a binary m-dimensional vector, then the norm of $\mathbf{A}$ is defined by

$$\| \mathbf{A} \| = \Sigma_{i=1}^{m} \; | A_i |. \tag{7.8}$$

Let **A** and **B** be binary m-dimensional vectors, then a third binary m-dimensional vector $\mathbf{A} \cap \mathbf{B}$ may be defined by

$$(\mathbf{A} \cap \mathbf{B}) = 1 \quad \text{if and only if} \quad A_i = 1 \text{ and } B_i = 1. \tag{7.9}$$

All *ART 1 learning* is gated by F_2 activity. That is, the bottom up [forward] and the top down [backward or feedback] adaptive weights w_{iJ} and W_{Ji} can change only when the Jth F_2 node is active. Both types of weights are functions of the F_1 vector **X**. Stated as a differential equation, the *top-down* or *feedback learning rule* is

$$\frac{d}{dt} W_{ji} = Y_j \, (X_i - W_{ji}), \tag{7.10}$$

where learning by W_{ji} is gated by Y_j. When the Y_j gate opens (i.e., when $Y_j > 0$), then learning begins and W_{ji} is attracted to X_i:

$$W_{ji} \rightarrow X_i. \tag{7.11}$$

In vector terms: if $Y_j > 0$, then $\mathbf{W}_j = (W_{j1}, ..., W_{jn})$ approaches $\mathbf{X} = (X_1, ..., X_n)$. Such a learning rule is termed *outstar learning rule* (Grossberg 1969). Initially, all W_{ji} are maximal, i.e.

$$W_{ji}(0) = 1. \tag{7.12}$$

Thus, with fast learning, where the adaptive weights fully converge to equilibrium values in response to each input pattern, the top-down [feedback] weight vector $\mathbf{W}_J$ is a binary vector at the start and end of each input presentation. By (7.3), (7.4), (7.9), (7.11) and (7.12), the *binary* F_1 *activity* [output] vector is given by

$$\mathbf{X} = \begin{cases} \mathbf{I} & \text{if } F_2 \text{ is inactive} \\ \mathbf{I} \cap \mathbf{W}_J & \text{if the jth } F_2 \text{ node is active.} \end{cases} \tag{7.13}$$

When F_2 node J is active, by (7.4) and (7.10) learning causes

$$\mathbf{W}_J \rightarrow \mathbf{I} \cap \mathbf{W}_J \text{ (old)} \tag{7.14}$$

In this *learning update rule* $\mathbf{W}_J$ (old) denotes $\mathbf{W}_J$ at the start of the current input presentation. By (7.11) and (7.13), $\mathbf{X}$ remains constant during learning, even though $|\mathbf{W}_J|$ may decrease. The first time an F_2 node J becomes active it is said to be *uncommitted*. Then, by (7.12) – (7.14)

$$\mathbf{W}_J \rightarrow \mathbf{I}. \tag{7.15}$$

The bottom up or forward weights have a slightly more complicated learning rule which leads to a similar, but normalized result. The combination with F_2 nodes which undergo cooperative and competitive interactions is called *competitive learning*. Initially all F_2 nodes are uncommitted. Forward weights w_{ij} in $F_1 \rightarrow F_2$ paths initially satisfy

$$w_{ij}(0) = \alpha_j, \tag{7.16}$$

where the parameters α_j are ordered according to $\alpha_1 > \alpha_2 > ... > \alpha_n$ for any admissible $F_0 \rightarrow F_1$ input $\mathbf{I}$.

Like the top-down weight vector $\mathbf{W}_J$, the bottom-up $F_1 \rightarrow F_2$ weight vector $\mathbf{w}_J = (w_{1J}, ..., w_{iJ}, ..., w_{nJ})$ also becomes proportional to the F_1 output vector $\mathbf{X}$ when the F_2 node J is active. But in addition the forward weights are scaled inversely to $||\mathbf{X}||$, so that

$$w_{iJ} \longrightarrow \frac{X_i}{\beta + ||\mathbf{X}||} \tag{7.17}$$

with $ß > 0$ [the small number ß is included to break ties]. This $F_1 \rightarrow F_2$ learning law (Carpenter and Grossberg 1987a) realizes a type of competition among the weights $\mathbf{w}_J$ adjacent to a given F_2 node J.

By (7.13), (7.14) and (7.17), during learning

$$\mathbf{w}_J \longrightarrow \frac{\mathbf{I} \cap \mathbf{W}_J \,(\text{old})}{\beta + \| \mathbf{I} \cap \mathbf{W}_J \,(\text{old}) \|}. \tag{7.18}$$

(7.18) establishes the update rule for forward weights. The w_{ij} initial values are required to be sufficiently small so that an input **I** which perfectly matches a previously learned vector $\mathbf{w}_J$ will select the F_2 node J rather than an uncommitted node. This is accomplished by assuming that

$$0 < \alpha_j = w_{ij}(0) < \frac{1}{\beta + \| \mathbf{I} \|} \tag{7.19}$$

for all $F_1 \rightarrow F_2$ inputs **I**. When **I** is first presented, $\mathbf{X} = \mathbf{I}$, so by (7.5), (7.14), (7.16), and (7.18), the $F_1 \rightarrow F_2$ input vector $T = (T_1, ..., T_m)$ obeys

$$T_j(\mathbf{I}) = \Sigma_{i=1}^{n} \; I_i \; w_{ij} = \begin{cases} \| \mathbf{I} \| \alpha_j & \text{if j is an uncommitted node} \\ \dfrac{\| \mathbf{I} \cap \mathbf{W}_j \|}{\beta + \| \mathbf{W}_j \|} & \text{if j is a committed node.} \end{cases} \tag{7.20}$$

(7.20) is termed the choice function in ART 1, where ß is the choice parameter and $ß \cong 0$. The limit $ß \rightarrow 0$ is called *conservative limit*, because small ß-values tend to minimize recoding during learning. If ß is taken so small then, among committed F_2 nodes, T_j is determined by the size $\| \mathbf{I} \cap \mathbf{W}_j \|$ relative to $\| \mathbf{W}_j \|$. Additionally, α_j values are taken to be so small that an uncommitted F_2 node will generate the maximum T_j value in (7.20) only if $\| \mathbf{I} \cap \mathbf{W}_j \| = 0$ for all committed nodes. Larger values of α_j and ß bias the system towards earlier selection of uncommitted nodes, when only poor matches are found among the committed nodes (for a more detailed discussion see Carpenter and Grossberg 1987a).

Rules for Search

It is important to note that ART 1 overcomes the stability-plasticity dilemma by accepting and adapting the prototype of a category [class] stored in F_2 only when the input pattern is sufficiently similar to it. In this case, the input pattern and the stored prototype are said to *resonate* [hence the term *resonance* theory]. When an input pattern fails to match any existing prototype [node] in F_2, a new category is formed (as in Hartigan's 1975, leader algorithm), with the input pattern as the prototype, using a previously uncommitted F_2 unit. If there are no such uncommitted nodes left, then a novel input pattern gives no response (see Hertz, Krogh and Palmer 1991).

A dimensionless parameter ρ with $0<\rho\leq 1$ which is termed *vigilance parameter* establishes a matching [similarity] criterion for deciding whether the similarity is good enough for the input pattern to be accepted as an example of the chosen prototype. The degree of match [similarity] between bottom-up input $\mathbf{I}$ and top-down expectation $\mathbf{W}_j$ is evaluated at the orienting subsystem of ART 1 (see Fig. 7.1) which measures whether prototype J adequately represents input pattern $\mathbf{I}$. A *reset* occurs when the match fails to meet the criterion established by the parameter ρ.

In fast-learning ART 1 with choice at F_2, the *search process* may be characterized by the following steps:

Step 1: Select one F_2 node J that maximizes T_j in (7.20), and read-out its top-down [feedback] weight vector $\mathbf{W}_J$.

Step 2: With J active, compare the F_1 output vector $\mathbf{X} = \mathbf{I} \cap \mathbf{W}_J$ with the $F_0 \rightarrow F_1$ input vector $\mathbf{I}$ at the orienting subsystem (see Fig. 7.1).

Step 3A: Suppose that $\mathbf{I} \cap \mathbf{W}_J$ fails to match $\mathbf{I}$ at the level required by the ρ-criterion, i.e. that

$$\|\mathbf{X}\| = \|\mathbf{I} \cap \mathbf{W}_J\| < \rho \|\mathbf{I}\|. \tag{7.21}$$

This mismatch causes the system to reset and inhibits the winning node J for the duration of the input interval during which $\mathbf{I}$ remains on. The index of the chosen prototype [F_2 node] is reset to the value corresponding to the next highest $F_1 \rightarrow F_2$ input $\mathbf{T}_j$. With the new node active, steps 2 and 3A are repeated until the chosen prototype satisfies the similarity [resonance] criterion (7.21).

Step 3B: Suppose that $\mathbf{I} \cap \mathbf{W}_J$ meets the similarity [match function] criterion, i.e.

$$\|\mathbf{X}\| = \|\mathbf{I} \cap \mathbf{W}_J\| \geq \rho \|\mathbf{I}\|, \tag{7.22}$$

then ART 1 search ends and the last chosen F_2 node J remains active until input $\mathbf{I}$ shuts off [or until ρ increases].

In this state, called *resonance*, both the feedforward ($F_1 \rightarrow F_2$) and the feedback ($F_2 \rightarrow F_1$) adaptive weights are updated if $\mathbf{I} \cap \mathbf{W}_J\,(\text{old}) \neq \mathbf{W}_J\,(\text{old})$. If ρ is chosen to be large [i.e. close to 1], the similarity condition becomes very stringent so that many finely divided categories [classes] are formed. A ρ–value close to zero gives a coarse categorization. The vigilance level can be changed during learning.

Finally, it is worth noting that ART 1 is exposed to discrete presentation intervals during which an input is constant and after which F_1 and F_2 activities are set to zero. Discrete presentation intervals are implemented by means of the F_1 and F_2 gain control signals (g_1, g_2). Gain signal g_2 is assumed (like g_1 in (7.2)) to be 0 if F_0 is inactive. When F_0 becomes active, g_2 and F_2 signal thresholds are assumed to lie in a range where the F_2 node which receives the largest input signal can become active.

7.3 The ARTMAP Neural Network Architecture

ARTMAP is a neural network architecture designed to solve supervised pattern recognition problems. The architecture is called ARTMAP because it maps input vectors in $\Re^n$ [such as feature vectors denoting spectral values of a pixel] to output vectors in $\Re^m$ [with m<n], representing predictions such as land use categories, where mapping is learned by example from pairs $\{\mathbf{A}^{(p)}, \mathbf{B}^{(p)}\}$ of sequentially presented input and output vectors p=1,2,3,... and $\mathbf{B}^{(p)}$ is the correct prediction given $\mathbf{A}^{(p)}$. Fig. 7.2 illustrates the main components of a binary ARTMAP system. The system incorporates two ART 1 modules, ART_a and ART_b. Indices a and b identify terms in the ART_a and ART_b modules, respectively. Thus, for example ρ_a and ρ_b denote the ART_a and ART_b vigilance [similarity] parameters, respectively.

During supervised learning ART_a and ART_b read vector inputs **A** and **B**. The ART_a complementing coding preprocessor transforms the vector $\mathbf{A}=(A_1, ..., A_{na})$ into the vector $\mathbf{I}^a= (\mathbf{A}, \mathbf{A}^C)$ at the ART_a field F_0^a, where $\mathbf{A}^C$ denotes the complement of **A**. The complement coded input $\mathbf{I}^a$ to the recognition system is the 2na-dimensionable vector

$$\mathbf{I}^a = (\mathbf{A}, \mathbf{A}^C) = (A_1, ..., A_{na}; \ A_1^C, ..., A_{na}^C), \tag{7.23}$$

where

$$A_i^C := 1 - A_i. \tag{7.24}$$

Complement coding achieves normalization while preserving amplitude information (see Carpenter, Grossberg and Reynolds 1991a). $\mathbf{I}^a$ is the input to the ARTa field F_1^a. Similarly, the input to the ARTb field F_1^b is the vector $\mathbf{I}^b$=($\mathbf{B}$, $\mathbf{B}^C$).

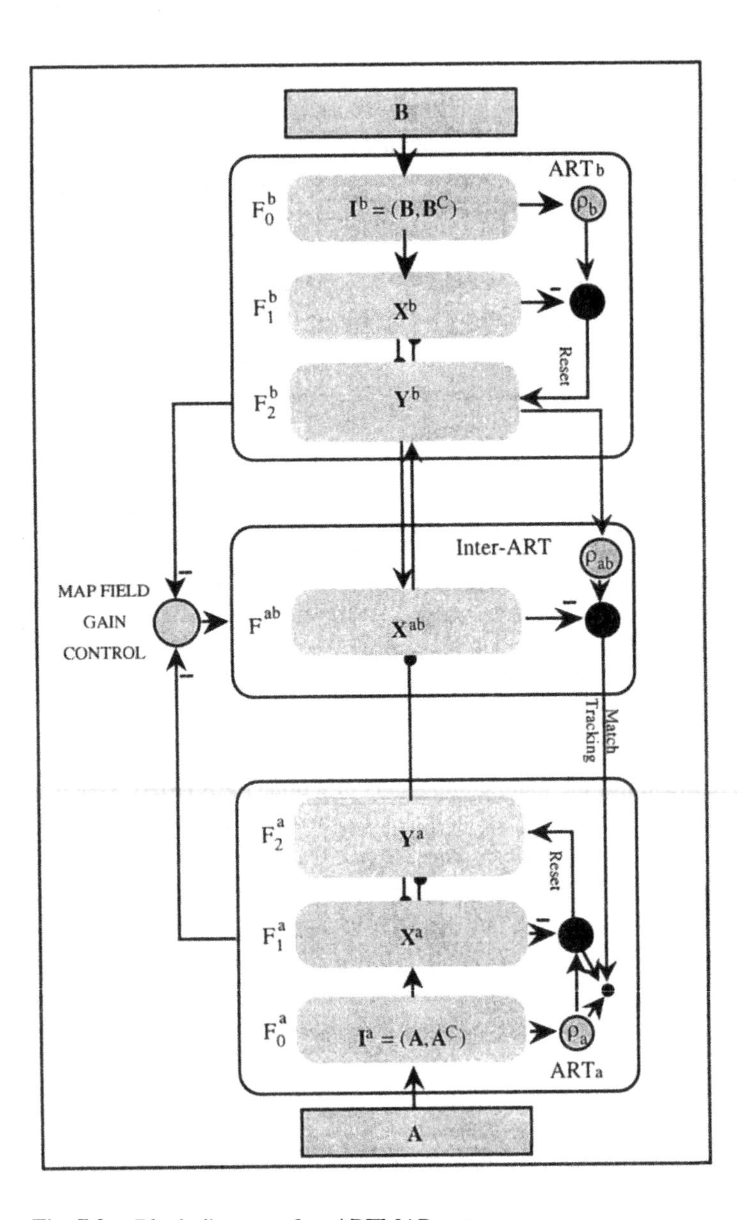

Fig. 7.2 Block diagram of an ARTMAP system

If ARTa and ARTb were disconnected, each module would self-organize category groupings for the separate input sets $\{\mathbf{A}^{(p)}\}$ and $\{\mathbf{B}^{(p)}\}$, respectively, as described in Section 7.2. In an ARTMAP architecture design, however, ARTa and ARTb are connected by an inter-ART module, including a map field that controls the learning of an associative map from ARTa recognition categories [i.e. compressed representations of classes of examplars $\mathbf{A}^{(p)}$] to ARTb recognition categories [i.e. compressed representations of classes of examplars $\mathbf{B}^{(p)}$]. Because the map field is the interface, where signals from F_2^a and F_2^b interact, it is denoted by F^{ab}. The nodes of F^{ab} have the same index j [j=1, ..., m_b] as the nodes of F_2^b because there is a one-to-one correspondence between these sets of nodes.

ARTa and ARTb operate as outlined in Section 7.2 with the following additions. First, the ARTa vigilance [similarity] parameter ρ_a can increase during inter-ART reset according to the match tracking rule. Second, the map field F^{ab} can prime ARTb. This means, if F^{ab} sends non uniform input to F_2^b in the absence of an $F_0^b \rightarrow F_1^b$ input $\mathbf{B}$, then F_2^b remains inactive. But as soon as an input arrives, F_2^b selects the node J receiving the largest $F^{ab} \rightarrow F_2^b$ input. Node J, in turn, sends to F_1^b the top-down input weight vector $\mathbf{W}_J^b$. Rules for the control strategy, called match tracking, are specified below (Carpenter, Grossberg and Reynolds 1991a).

Let $\mathbf{X}^a = (X_1^a, ..., X_{na}^a)$ denote the F_1^a output vector and $\mathbf{Y}^a = (Y_1^a, ..., Y_{ma}^a)$ the F_2^a output vector. Similarly, let denote $\mathbf{X}^b = (X_1^b, ..., X_{nb}^b)$ the F_1^b output vector and $\mathbf{Y}^b = (Y_1^b, ..., Y_{mb}^b)$ the F_2^b output vector. The map field F^{ab} has m_b nodes and binary output vector $\mathbf{X}^{ab}$. Vectors $\mathbf{X}^a$, $\mathbf{Y}^a$, $\mathbf{X}^b$, $\mathbf{Y}^b$ and $\mathbf{X}^{ab}$ are set to the zero vector, $\mathbf{0}$, between input presentations.

The $F_2^a \rightarrow F^{ab}$ adaptive weights z_{kj} with k=1, ..., m_a and j=1, ..., m_b obey an outstar learning law similar to that governing the $F_2^b \rightarrow F_1^b$ weights, namely

$$\frac{d}{dt} z_{kj} = \mathbf{Y}_k^a (\mathbf{X}_k^{ab} - z_{kj}). \tag{7.25}$$

Each vector $(z_{k1}, ..., z_{kmb})$ is denoted by $\mathbf{z}_k$. According to the learning rule established by (7.25), the $F_2^a \rightarrow F^{ab}$ weight vector $\mathbf{z}_k$ approaches the map field F^{ab} activity vector $\mathbf{X}^{ab}$ if the k-th F_2^a node is active. Otherwise $\mathbf{z}_k$ remains constant. If node k has not yet learned to make a prediction, all weights z_{kj} are set equal to 1, using an assumption, analogous to Equation (7.12), i.e. $z_{kj}(0)=1$ for k=1, ..., m_a, and j=1, ..., m_b.

During resonance with ARTa category k active, $\mathbf{z}_k \rightarrow \mathbf{X}^{ab}$. In fast learning, once k learns to predict the ARTb category J, that association is permanent [i.e. $z_{kJ} = 1$ and $z_{kj} = 0$ with $j \neq J$ for all time]. The F^{ab} output vector $\mathbf{X}^{ab}$ obeys

$$\mathbf{X}^{ab} = \begin{cases} \mathbf{Y}^b \cap \mathbf{z}_k & \text{if the k-th } F_2^a \text{ node is active and } F_2^b \text{ is active} \\ \mathbf{z}_k & \text{if the k-th } F_2^a \text{ node is active and } F_2^b \text{ is inactive} \\ \mathbf{Y}^b & \text{if } F_2^a \text{ is inactive and } F_2^b \text{ is active} \\ \mathbf{0} & \text{if } F_2^a \text{ is inactive and } F_2^b \text{ is inactive.} \end{cases} \tag{7.26}$$

When ARTa makes a prediction that is incompatible with the actual ARTb input, i.e. $\mathbf{z}_k$ is disconfirmed by $\mathbf{Y}^b$, then this mismatch triggers on ARTa search for a new category as follows. At the start of each input presentation the ARTa vigilance [similarity] parameter ρ_a equals a baseline vigilance $\overline{\rho_a}$. The map field vigilance parameter is ρ_{ab}. If a mismatch at F^{ab} occurs, i.e. if

$$\| \mathbf{X}^{ab} \| < \rho_{ab} \| \mathbf{I}^b \|, \tag{7.27}$$

then match tracking is triggered to search a new F_2^a node. Match tracking starts a cycle of ρ_a adjustment and increases ρ_a until it is slightly higher than the F_1^a match value $\| \mathbf{A} \cap \mathbf{W}_k^a \| \| \mathbf{I}^a \|^{-1}$, where $\mathbf{W}_k^a$ denotes the top-down $F_2^a \rightarrow F_1^a$ ARTa weight vector $(\mathbf{W}_1^a, \ldots, \mathbf{W}_{na}^a)$. Then

$$\| \mathbf{X}^a \| = \| \mathbf{I}^a \cap \mathbf{W}_k^a \| < \rho_a \| \mathbf{I}^a \| \tag{7.28}$$

where $\mathbf{I}^a$ is the current ARTa input vector and k is the index of the active F_2^a node. When this occurs, ARTa search leads either to ARTMAP resonance, where a newly chosen F_2^a node K satisfies both the ARTa matching criterion (see also Equation (7.21)):

$$\| \mathbf{X}^a \| = \| \mathbf{I}^a \cap \mathbf{W}_K^a \| \geq \rho_a \| \mathbf{I}^a \| \tag{7.29}$$

and the map field matching criterion:

$$\| \mathbf{X}^{ab} \| = \| \mathbf{Y}^b \cap \mathbf{z}_K \| \geq \rho_{ab} \| \mathbf{Y}^b \| \tag{7.30}$$

or, if no such node K exists, to the shut-down of F_2^a for the remainder of the input presentation (Carpenter, Grossberg and Ross 1993).

7.4 Generalization to Fuzzy ARTMAP

Fuzzy ARTMAP has been proposed by Carpenter, Grossberg and Rosen (1991b) as a direct generalization of ARTMAP for supervised learning of recognition categories and multidimensional maps in response to arbitrary sequences of continuous-valued [and binary] patterns, not necessarily interpreted as fuzzy set of features. The generalization to learning continuous and binary input patterns is achieved by using fuzzy set operations, rather than standard binary set theory operations (see Zadeh 1965). Fig. 7.3 summarizes how the crisp logical ARTMAP operations of category choice, matching and learning translate into fuzzy ART operations when the crisp [non-fuzzy or hard] intersection operator ($\cap$) of ARTMAP is replaced by the fuzzy intersection or [component-wise] minimum operator ($\wedge$). Due to the close formal homology between ARTMAP and fuzzy ARTMAP operations (as illustrated in Fig. 7.3), there is no need to describe fuzzy ARTMAP in detail here, but for a better understanding it is important to stress differences between this and the ARTMAP approach.

Fuzzy ARTMAP in its most general form inherits the architecture outlined in Fig. 7.2 and employs two fuzzy ART modules as substitutes for the ART 1 subsystems. It is noteworthy that fuzzy ART reduces to ART 1 in response to binary input vectors (Carpenter, Grossberg and Ross 1993). Associated with each F_2^a [F_2^b] node k=1, ..., m_a [j=1, ..., m_b] is a vector $\mathbf{W}_k^a$ [$\mathbf{W}_j^b$] of adaptive weights which subsumes both the bottom-up and top-down weight vectors of ART 1.
Fuzzy ARTMAP dynamics are determined by a choice parameter $\beta > 0$, a learning parameter $\gamma \in [0,1]$, and three vigilance ([similarity] parameters: the ARTa vigilance parameter ρ_a, the ARTb vigilance parameter ρ_b and the map field vigilance parameter ρ_{ab} with $\rho_a, \rho_b, \rho_{ab} \in [0,1]$. The choice functions $T_k(\mathbf{A})$ and $T_j(\mathbf{B})$ are shown in Fig. 7.3, where the fuzzy intersection ($\wedge$) for any n-dimensional vectors $\mathbf{S}=(S_1, ..., S_n)$ and $T=(T_1, ..., T_n)$ is defined by

$$(\mathbf{S} \wedge \mathbf{T})_l = \min_l (S_l, T_l) \tag{7.31}$$

The fuzzy choice functions $T_k(\mathbf{A})$ and $T_j(\mathbf{B})$ (see Fig. 7.3) can be interpreted as fuzzy membership of the input **A** in the k-th category and the input **B** in the j-th category, respectively. In the conservative limit (i.e. $\beta \to 0$) the choice function $T_k(\mathbf{A})$ primarily reflects the degree to which the weight vector $\mathbf{W}_k^a$ is a fuzzy subset of the input vector **A**. If

$$\frac{\|\mathbf{I}_l^a \wedge \mathbf{W}_k^a\|}{\|\mathbf{W}_k^a\|} = 1, \tag{7.32}$$

then $\mathbf{W}_k^a$ is a fuzzy subset of $\mathbf{I}^a$ and category k is said to be a fuzzy subset choice for input $\mathbf{I}^a$.

	Binary ARTMAP	Fuzzy ARTMAP
ARTa Category Choice [β choice parameter]	$T_k(I^a) = \dfrac{\lVert \mathbf{I}^a \cap \mathbf{W}_k^a \rVert}{\beta + \lVert \mathbf{W}_k^a \rVert}$	$T_k(I^a) = \dfrac{\lVert \mathbf{I}^a \wedge \mathbf{W}_k^a \rVert}{\beta + \lVert \mathbf{W}_k^a \rVert}$
ARTb Category Choice [β choice parameter]	$T_j(I^b) = \dfrac{\lVert \mathbf{I}^b \cap \mathbf{W}_j^b \rVert}{\beta + \lVert \mathbf{W}_j^b \rVert}$	$T_j(I^b) = \dfrac{\lVert \mathbf{I}^b \wedge \mathbf{W}_j^b \rVert}{\beta + \lVert \mathbf{W}_j^b \rVert}$
ARTa Matching Criterion [ρ_a ARTa vigilance parameter]	$\lVert I^a \cap \mathbf{W}_K^a \rVert \geq \rho_a \lVert A \rVert$	$\lVert I^a \wedge \mathbf{W}_K^a \rVert \geq \rho_a \lVert A \rVert$
ARTb Matching Criterion [ρ_b ARTb vigilance parameter]	$\lVert I^b \cap \mathbf{W}_J^b \rVert \geq \rho_b \lVert B \rVert$	$\lVert I^a \wedge \mathbf{W}_J^b \rVert \geq \rho_b \lVert B \rVert$
Map Field F^{ab} Matching Criterion [ρ_{ab} Map field vigilance parameter]	$\lVert X^{ab} \rVert = \lVert Y^b \cap z_K \rVert \geq \rho_{ab} \lVert Y^b \rVert$	$\lvert X^{ab} \rVert = \lVert Y^b \wedge z_K \rVert \geq \rho_{ab} \lVert Y^b \rVert$
ARTa $F_2^a \rightarrow F_1^a$ **Learning Weight Updates** [γ learning parameter]	$\mathbf{W}_K^a (\text{new}) = \gamma (A \cap \mathbf{W}_K^a (\text{old})) + (1-\gamma)$ $\mathbf{W}_K^a (\text{old})$	$\mathbf{W}_K^a (\text{new}) = \gamma (A \square \mathbf{W}_K^a (\text{old})) + (1-\gamma)$ $\mathbf{W}_K^a (\text{old})$
ARTb $F_2^b \rightarrow F_1^b$ **Learning Weight Updates** [γ learning parameter]	$\mathbf{W}_J^b (\text{new}) = \gamma (B \cap \mathbf{W}_J^b (\text{old})) + (1-\gamma)$ $\mathbf{W}_J^b (\text{old})$	$\mathbf{W}_J^b (\text{new}) = \gamma (B \wedge \mathbf{W}_J^b (\text{old})) + (1-\gamma)$ $\mathbf{W}_J^b (\text{old})$

Fig. 7.3 The fuzzy ARTMAP classifier in comparison with the binary ARTMAP classifier: A simplified ARTMAP architecture

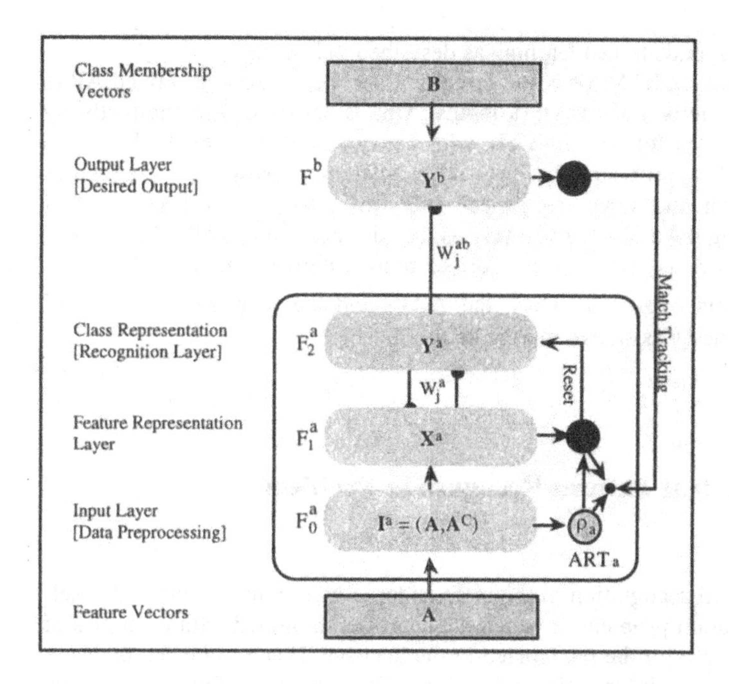

Fig. 7.4 The fuzzy ARTMAP classifier: A simplified ARTMAP architecture

When a fuzzy subset exists, it is always selected over other choices. The same holds true for $T_j(\mathbf{I}^b)$. (Carpenter et al. 1992). Resonance depends on the degree to which $\mathbf{I}^a[\mathbf{I}^b]$ is a fuzzy set of $\mathbf{W}_K^a$ [$\mathbf{W}_K^b$], by the matching criteria [or functions] outlined in Fig. 7.3. The close linkage between fuzzy subsethood and ART choice, matching and learning forms the foundations of the computational features of fuzzy ARTMAP (Carpenter et al. 1992). Especially if category K is a fuzzy subset ART_a choice, then the ART_a match function value ρ_a is given by

$$\rho_a = \frac{\|\mathbf{I}^a \wedge \mathbf{W}_K^a\|}{\|\mathbf{I}^a\|} = \frac{\|\mathbf{W}_K^a\|}{\|\mathbf{I}^a\|} . \tag{7.33}$$

Once search ends, the ART_a weight vector $\mathbf{W}_K^a$ is updated according to the equation

$$\mathbf{W}_K^a\,(\text{new}) = \gamma\,(\mathbf{A} \wedge \mathbf{W}_K^a\,(\text{old})) + (1-\gamma)\;\mathbf{W}_K^a\,(\text{old}) \tag{7.34}$$

and similarly the ART_b weight vector $\mathbf{W}_J^b$:

$$\mathbf{W}_J^b\,(\text{new}) = \gamma\,(\mathbf{B} \wedge \mathbf{W}_J^b\,(\text{old})) + (1-\gamma)\;\mathbf{W}_J^b\,(\text{old}) \tag{7.35}$$

where $\gamma = 1$ corresponds to fast learning as described in Fig. 7.3.

The aim of fuzzy ARTMAP is to correctly associate continuous valued ART_a inputs with continuous valued ART_b inputs. This is accomplished indirectly by associating categories formed in ART_a with categories formed in ART_b. For a pattern classification problem, the desired association is between a continuous valued input vector and some categorical code which takes on a discrete set of values representing the a priori given classes. In this situation the ART_b network is not needed because the internal categorical representation which ART_b would learn already exists explicitly. Thus, the ART_b and the map field F^{ab} can be replaced by a single F_b as shown in Fig. 7.4.

7.5 The Spectral Pattern Recognition Problem

The spectral pattern recognition problem considered here is the supervised pixel-by-pixel classification problem in which the classifier is trained with examples of the classes [categories] to be recognized in the data set. This is achieved by using limited ground survey information which specifies where examples of specific categories are to be found in the imagery. Such ground truth information has been gathered on sites which are well representative of the much larger area analysed from space. The image data set consists of 2,460 pixels [resolution cells] selected from a Landsat-5 Thematic Mapper [TM] scene [270x360 pixels] from the city of Vienna and its northern surroundings [observation date: June 5, 1985; location of the centre: 16°23' E, 48°14' N; TM Quarter Scene 190-026/4]. The six Landsat TM spectral bands used are blue [SB1], green [SB2], red [SB3], near IR [SB4], mid IR [SB5] and mid IR [SB7], excluding the thermal band with only a 120m ground resolution. Thus, each TM pixel represents a ground area of 30x30 m^2 and has six spectral band values ranging over 256 digital numbers [8 bits].

The purpose of the multispectral classification task at hand is to distinguish between the eight categories of urban land use listed in Table 7.1. The categories chosen are meaningful to photo-interpreters and land use managers, but are not necessarily spectrally homogeneous. This prediction problem, used to evaluate the performance of fuzzy ARTMAP in a real world context, is challenging. The pixel-based remotely sensed spectral band values are noisy and sometimes unreliable. The number of training sites is small relative to the number of land use categories [one-site training case]. Some of the urban land use classes are sparsely distributed in the image. It is reported that conventional statistical classifiers, such as the Gaussian maximum likelihood classifier, have failed to discriminate spectrally inhomogeneous classes such as C6 (see, e.g., Hepner et al. 1990). Thus, there is evidently a need for new more powerful tools (Barnsley 1993).

Ideally, the ground truth at every pixel of the scene should be known. Since this is impractical, one training site was chosen for each of the eight above mentioned land use categories. The training sites vary between 154 pixels [category: suburban] and 602 pixels [category: woodland and public gardens with trees]. The above mentioned six TM bands provide the data set input for each pixel, with values scaled to the interval [0,1]. This approach resulted in a data base consisting of 2,460 pixels [about 2.5 percent of all the pixels in the scene] that are described by six-dimensional feature vectors, each tagged with its correct category membership. The set was divided into a training set [two thirds of the training site pixels] and a testing set by stratified random sampling, stratified in terms of the eight categories. Pixels from the testing set were not used during network training [parameter estimation] and served only to evaluate out-of-sample test [prediction, generalization] performance accuracy when the trained classifier is presented with novel data. The goal is to predict the correct land use category for the test sample of pixels.

Table 7.1 Categories used for classification and number of training/testing pixels

Category	Description of the Category	Pixels	
		Training	Testing
C1	Mixed grass and arable farmland	167	83
C2	Vineyards and areas with low vegetation cover	285	142
C3	Asphalt and concrete surfaces	128	64
C4	Woodland and public gardens with trees	402	200
C5	Low density residential and industrial areas	102	52
C6	Densely built up residential areas [urban]	296	148
C7	Water courses	153	77
C8	Stagnant water bodies	107	54
Total Number of Pixels for Training and Testing		1.640	820

A good classifier is one which, after training with the training set of pixels, is able to predict pixel assignments from the remotely sensed data over much wider areas of territory without the need for further ground survey (see Wilkinson, Fierens and Kanellopoulos 1995). The performance of any classifier is thus dependent upon three factors: i) the adequacy of the training set of pixels and, therefore, the choice of training sites; ii) the in-sample performance of the classifier, and; iii) the out-of-sample or generalization performance of the trained classifier. Of these three factors the first is often outside the control of the data analyst, and thus outside of the scope of this chapter.

7.6 Fuzzy ARTMAP Simulations and Classification Results

In this real world setting, fuzzy ARTMAP performance is examined and compared with that of the multi-layer perceptron, as well as the conventional maximum likelihood classifier. In-sample and out-of sample performance is measured in terms of the fraction of the total number of correctly classified pixels [i. e. the sum of the elements along the main diagonal of the classification error matrix].

During training and testing, a given pixel provides an ART_a input $\mathbf{A}=(A_1, A_2, A_3, A_4, A_5, A_6)$ where A_1 is blue, A_2 green, A_3 is red, A_4 near infrared, A_5 and A_6 mid infrared [1.55–1.75 μm and 2.08–2.35 μm, respectively] spectral band values measured at each pixel. The corresponding ART_b input vector **B** represents the correct land use category of the pixel's site:

$$
\mathbf{B} = \begin{cases}
(1,0,0,0,0,0,0,0) & \text{for mixed grass and arable farmland; category 1} \\
(0,1,0,0,0,0,0,0) & \text{for vineyards and areas with low vegetation cover; category 2} \\
(0,0,1,0,0,0,0,0) & \text{for asphalt and concrete surfaces; category 3} \\
(0,0,0,1,0,0,0,0) & \text{for woodland and public gardens with trees; category 4} \\
(0,0,0,0,1,0,0,0) & \text{for low density residential and industrial areas; category 5} \\
(0,0,0,0,0,1,0,0) & \text{for densely built-up residential areas; category 6} \\
(0,0,0,0,0,0,1,0) & \text{for water courses; category 7} \\
(0,0,0,0,0,0,0,1) & \text{for stagnant water bodies; category 8.}
\end{cases}
$$

During training vector **B** informs the fuzzy ARTMAP classifier of the land use category to which the pixel belongs. This supervised learning process allows adaptive weights to encode the correct associations between **A** and **B**. The remote sensing problem described in Section 7.5 requires a trained fuzzy ARTMAP network to predict the land use category of the test set pixels, given six spectral band values measured at each pixel.

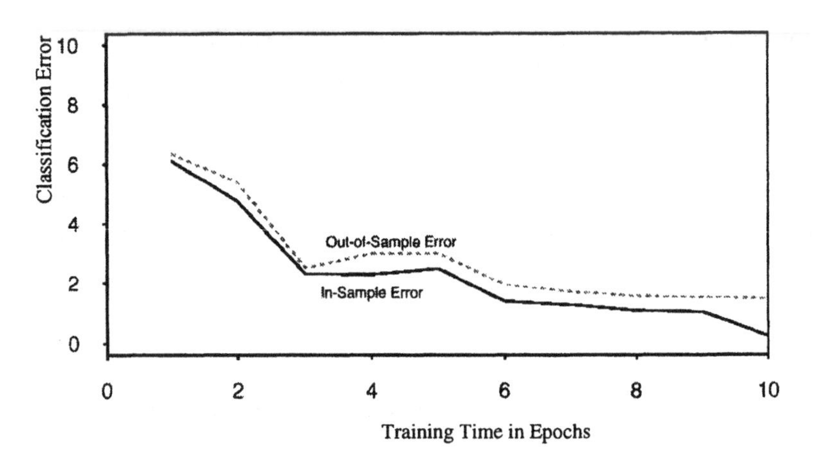

Fig. 7.5 In-sample and out-of-sample classification error during training [β=0.001, γ=1.0, ρ_a=0.001]

Fuzzy ARTMAP is trained incrementally, with each spectral band vector **A** presented just once. Following a search, if necessary, the classifier selects an ART_a category by activating an F_2^a node K for the chosen pixel, and learns to associate category K with the ART_b land use category of the pixel. With fast learning [γ=1], the class prediction of each ART_a category K is permanent. If some input **A** with a different class prediction later chooses this category, match tracking will raise vigilance ρ_a just enough to trigger a search for a different ART_a category. If the finite input set is presented repeatedly, then all training set inputs learn to predict with 100% classification accuracy, but start to fit noise present in the remotely sensed spectral band values.

All the simulations described below use the simplified fuzzy ARTMAP architecture, outlined in Fig. 7.3, with only three parameters: a choice parameter β>0, the learning parameter γ=1 [fast learning], and an ART_a vigilance parameter $\rho_a \in [0,1]$. In each simulation, the training data set represents 1,640 pixels and the testing data set 820 pixels. Since input order may affect in-sample and out-of-sample performance, fuzzy ARTMAP was run with five different random orderings of the training and test sets. All simulations were carried out at the Department of Economic Geography [WU-Wien] on a SunSPARCserver 10-GS with 128 MB RAM.

Table 7.2 summarizes out-of-sample performance [measured in terms of classification accuracy] on 15 simulations, along with the number of ART_a categories generated and number of epochs needed to reach asymptotic training set performance, i.e. about 100% in-sample classification accuracy. It is shown in Fig. 7.5 how in-sample and out-of-sample performance changes according to the number of training epochs of fuzzy ARTMAP. Each run had a different, randomly chosen, presentation order for the 1,640 training and 820 testing vectors. The choice parameter β was set, first, near the conservative limit at value β=0.001, and then at the higher values of β=0.1 and β=1.0. These β-value inputs were repeatedly presented in a given random order until 100% training classification accuracy was reached.

This required six to eight epochs in the cases of β=0.001 and β=0.1, while for β=1.0 eight to ten epochs were necessary. There seems to be a tendency for the number of epochs needed for 100% training set performance to increase with higher β-values. All simulations used fast learning [γ=1.0], which generates a distinct ART_a category structure for each input ordering. The number of F_2^a nodes ranged from 116 to 148 in the case of β=0.001, 115 to 127 in the case of β=0.1, and 202 to 236 in the case of β=1.0. This tendency for the number of ART_a categories to rise with increasing β-values and increasing training time is illustrated in Fig. 7.6. All simulations used ρ_a= 0.0, which tends to minimize the number of F_2^a nodes compared with higher ρ_a-values, not shown in Table 7.2. The best average result [over five independent simulation runs] was obtained with β=0.01 and 6.5 epoch training [99.29% classification accuracy]. All the 15 individual simulation runs reached an out-of-sample performance close to 100% [range: 98.40 to 99.90%].

Table 7.2 Fuzzy ARTMAP simulations of the remote sensing classification problem: The effect of variations in choice parameter β [ρ_a=0.0]

Choice Parameter β	Out-of-Sample Performance	Number of F_a^2 Nodes	Number of Epochs
β = 0.001			
Run 1	98.54	125	6
Run 2	99.26	116	8
Run 3	99.51	121	6
Run 4	99.39	126	6
Run 5	99.75	148	7
Average	99.29	127	6.5
β = 0.1			
Run 1	99.26	126	7
Run 2	99.90	115	6
Run 3	99.36	115	7
Run 4	99.51	124	7
Run 5	99.26	127	7
Average	99.26	121	7
β = 1.0			
Run 1	99.98	218	10
Run 2	98.17	202	8
Run 3	98.90	212	8
Run 4	98.50	236	10
Run 5	98.40	232	10
Average	98.75	220	9

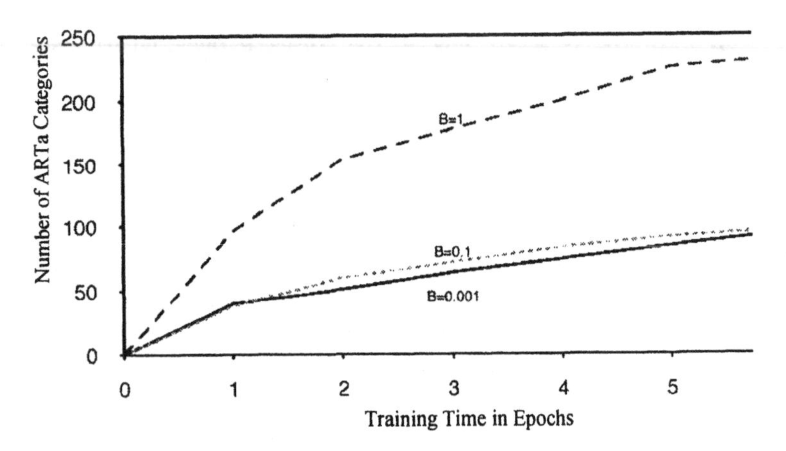

Fig. 7.6 Effect of choice parameter β on the number [m_a] of ARTa categories [γ=1.0, ρ_a=0.001]

Table 7.3 shows how in-sample and out-of-sample performance changes depending on the number of F_2^a nodes with ρ_a=0.95, 0.75, 0.50 and 0.0. In these simulations, learning is incremental, with each input presented only once [in ART terminology: one epoch training]. The choice parameter was set to β=0.001. The best overall results, in terms of average in-sample and out-of-sample performance were obtained with an ART_a vigilance close to one [96.36% and 95.82%, respectively]. For ρ_a=0.0 the in-sample and out-of-sample performances decline to 91.69% and 91.06%, respectively. But the runs with ρ_a= 0.0 use far fewer ART_a categories [32 to 44] compared to ρ_a= 0.95 [276 to 298 ART_a categories], and generate stable performance over the five runs. Increasing vigilance creates more ART_a categories. One final comment worth making is that most fuzzy ARTMAP learning occurs in the first epoch, with the test set performance on systems trained for one epoch typically over 92% that of systems exposed to inputs for six to eight epochs (compare Table 7.3 with Table 7.2).

Table 7.3 Fuzzy ARTMAP simulations of the remote sensing classification problem: The effect of variations in vigilance ρ_a (β = 0.001)

Vigilance [Similarity] Parameter ρ_a	In-Sample Performance	Out-of-Sample Performance	Number of F_a^2 Nodes
$\rho_a = 0.95$			
Run 1	97.01	96.20	285
Run 2	97.00	96.20	298
Run 3	96.15	95.60	276
Run 4	96.21	95.36	276
Run 5	95.06	94.39	286
Average	96.36	95.82	284
$\rho_a = 0.75$			
Run 1	93.00	92.00	52
Run 2	92.26	93.29	47
Run 3	91.82	90.00	42
Run 4	93.00	93.04	53
Run 5	90.31	91.83	53
Average	92.08	92.03	50
$\rho_a = 0.50$			
Run 1	92.20	91.40	43
Run 2	90.20	89.51	43
Run 3	94.45	94.76	44
Run 4	93.35	93.42	43
Run 5	92.98	93.90	45
Average	92.62	92.59	44
$\rho_a = 0.0$			
Run 1	90.70	90.60	35
Run 2	92.26	91.22	44
Run 3	90.97	90.30	34
Run 4	91.95	90,73	40
Run 5	92.56	92.44	32
Average	91.69	91.06	37

Table 7.4 summarizes the results of the third set of fuzzy ARTMAP simulations carried out, in terms of both in-sample and out-of-sample performance along with the number of F_2^a nodes. The choice parameter β was set near the conservative limit at value β=0.001 and ARTa vigilance at ρ_a= 0.0. Training lasted for one epoch only. As training size increases from 164 to 164,000 pixel vectors both in-sample and out-of-sample performances increase, but so does the number of ARTa category nodes. In-sample classification accuracy increases from 83.2% to 99.3%, and out-of-sample classification accuracy from 80.1% to 99.2%, while the number of ARTa category nodes increases from 19 to 225.

Table 7.4 Fuzzy ARTMAP simulations of the remote sensing classification problem: The effect of variations in training size (ρ_a= 0.0, β= 0.001)

Number of Training Pixels	In-Sample Performance	Out-of-Sample Performance	Number of F_a^2 Nodes
164	83.2	80.1	19
1,640	93.0	92.0	33
16,400	99.3	99.2	135
164,000	99.3	99.2	225

Each category node K requires six learned weights $\mathbf{w}_K^a$ in ARTa. One epoch training on 164 training pixels creates 19 ARTa categories, and so uses 72 ARTa adaptive weights to achieve 80.1% out-of-sample classification accuracy [820 test pixels], while one epoch training on 164,000 pixels requires 225 ARTa categories and, thus, 1,350 ARTa adaptive weights to arrive at an out-of-sample performance of 99.2%. Evidently, the fuzzy ARTMAP classifier becomes arbitrarily accurate provided the number of F_2^a nodes increases as needed.

Finally, fuzzy ARTMAP performance is compared with that of a multi-layer perceptron classifier as developed and implemented in Fischer et al. (1994), using the same training and testing set data. Table 7.5 summarizes the results of the comparison of the two neural classifers in terms of the in-sample and out-of-sample classification accuracies along with the number of epochs [i.e. one pass through the training data set] and the number of hidden units/ARTa category nodes [a hidden unit is somewhat analogous to an ARTa category for purposes of comparison] to reach asymptotic convergence, and the number of adaptive weight parameters. The fuzzy ARTMAP classifier has been designed with the following specifications: choice parameter near the conservative limit at value β=0.001, learning parameter γ=1.0, constant ARTa vigilance ρ_a=0.0, repeated presentation of inputs in a given order until 100% training set performance was reached. Stability and match tracking allow fuzzy ARTMAP to automatically construct as many ARTa categories as are needed to learn any consistent training set to 100% classification accuracy. The multi-layer perceptron classifier is a pruned, one hidden layer feedforward network with 14 logistic hidden units and eight softmax output units, using an epoch-based stochastic version of the backpropagation algorithm [epoch size: three training vectors, no momentum update, learning parameter γ=0.8]. The Gaussian maximum likelihood classifier based on parametric density estimation by maximum likelihood was chosen because it represents a widely used standard for comparison that yields minimum total classification error for Gaussian class distributions.

The fuzzy ARTMAP classifier has an outstanding out-of-sample classification accuracy of 99.26% on the 820 pixels testing data set. Thus the error rate [0.74%] is less than 1/15 that of the multi-layer perceptron and 1/20 that of the Gaussian maximum likelihood classifier. A careful inspection of the classification error [confusion] matrices [see Appendix] shows that there is however some confusion between certain land use categories – the densely built-up residential areas and water courses – by both the multi-layer perceptron and the Gaussian maximum likelihood classifiers, which is peculiar. The water body in this case is the river Danube that flows through the city and is surrounded by densely built up areas. The confusion could be caused by the 'boundary problem' where there are mixed pixels at the boundary. The fuzzy ARTMAP neural network approach evidently more easily accommodates a heterogeneous class label such as *densely built-up residential areas* to produce a visually and numerically correct map, even with smaller numbers of training pixels (see Fig. 7.7).

Table 7.5 Performance of fuzzy ARTMAP simulations of the remote sensing classification problem: Comparison with Multilayer Perceptron and the Gaussian Maximum Likelihood classifier

Classifier	Epochs[1]	Hidden Units/ ARTa Categories	Adaptive Weight Parameters	In-Sample Classification Accuracy	Out-of-Sample Classification Accuracy
Fuzzy ARTMAP	8	116	812	100.00	99.26
Multi-Layer Perceptron	92	14	196	92.13	89.76
Gaussian Maximum Likelihood	-	-	-	90.85	85.24

[1]one pass through the training data set

Fuzzy ARTMAP: $\beta = 00.1$, $\gamma = 1.0$, $\rho_a = 0.0$, asymptotic training

Multi-Layer Perceptron: logistic hidden unit activation, softmax output unit activation, network pruning, epoch based stochastic version of backpropagation with epoch size of three, learning rate $\gamma = 0.8$

The primary computational difference between the fuzzy ARTMAP and the multi-layer perceptron algorithms is speed. The backpropagation approach to neural network training is extremely computation-intensive, taking about one order of magnitude longer than the fuzzy ARTMAP, when implemented on a serial workstation. Although this situation may be alleviated with other, more efficient training algorithms and parallel implementation, it remains a serious drawback to the routine use of multi-layer perceptron classifiers. Finally, it should be mentioned that in terms of the total number of pathways [i.e. number of weight parameters] needed for the best performance, the multilayer perceptron classifier is superior to fuzzy ARTMAP, but at the higher computation costs and lower classification accuracies mentioned above.

Fig. 7.7 The fuzzy ARTMAP classified image

7.7 Summary and Conclusions

Classification of terrain cover from satellite imagery represents an area of considerable current interest. Satellite sensors record data in a variety of spectral channels and at a variety of ground resolutions. The analysis of remotely sensed data is usually carried out by machine-oriented pattern recognition techniques, the most widely used of which, assuming Gaussian distribution of the data, is classification based on maximum likelihood. In this chapter, we have compared the performance of fuzzy ARTMAP with an error-based learning system based on the multi-layer perceptron and Gaussian maximum-likelihood classifier as a conventional statistical benchmark on the same database. Both neural network classifiers outperform the conventional classifier in terms of map user, map producer and total classification accuracies. The fuzzy ARTMAP simulations led to by far the best out-of-sample classification accuracies, very close to maximum performance.

It was evident that the fuzzy ARTMAP classifier was able to accommodate more easily a heterogenenous class label, such as densely built-up residential areas, to produce a visually and numerically correct urban land use map, even with a smaller number of training pixels. The Gaussian maximum likelihood classifier tends to be sensitive to the purity of land use category signatures and performs poorly if they are not pure.

The study shows that the fuzzy ARTMAP classifier is a powerful tool for remotely sensed image classification. Even one epoch of fuzzy ARTMAP training yields close to maximum performance. The specific features of ART, such as speed and incremental learning, may mean that the fuzzy ARTMAP multispectral image classifier has the potential to become a standard tool in remote sensing, especially for data from future multichannel satellites such as the 224 channel Airborne Visible and Infrared Imaging Spectrometer [AVIRIS], and for the classification of multi-data and multi-temporal imagery, or for extending the same classification to different images. To conclude, we should like to mention that the classfier does however lead to crisp rather than fuzzy classifications, thus losing some of the attractivity of *fuzzy* pattern recognition systems. This is certainly one area in which the classifier could be further improved.

Acknowledgements: The authors wish to thank Professor Karl Kraus of the Department of Photogrammetric Engineering and Remote Sensing, Vienna Technical University, for his assistance in supplying the remote sensing data used in this study. The work was funded by the Austrian Ministry for Science, Traffic and Art [contract no. EZ 308.937/2-W/3/95]. The authors also like to thank Petra Staufer [Wirtschaftsuniversität Wien] for her help.

Endnote

1 This chapter is a reprint from Fischer M.M. and Getis A. (eds.) (1997): *Recent Developments in Spatial Analysis – Spatial Statistics, Behavioural Modelling, and Computational Intelligence.* Springer, Berlin, Heidelberg, New York, pp. 304-335

Appendix A
In-Sample and Out-of-Sample Classification Error Matrices of the Classifiers

An error matrix is a square array of numbers set out in rows and columns which expresses the number of pixels assigned to a particular category relative to the actual category as verified by some reference [ground truth] data. The rows represent the reference data, the columns indicate the categorization generated. It is important to note that differences between the map classification and reference data might be not only due to classification errors. Other possible sources of error include errors in interpretation and delineation of the reference data, changes in land use occurring between the collection of the remotely sensed data and reference data [temporal error], and variation in classification of the reference data due to inconsistencies in human interpretation.

Table A.1 In-sample performance: Classification error matrices

(a) Fuzzy ARTMAP

Ground Truth Categories	Classifier's Categories								
	C1	C2	C3	C4	C5	C6	C7	C8	Total
C1	167	0	0	0	0	0	0	0	167
C2	0	285	0	0	0	0	0	0	285
C3	0	0	128	0	0	0	0	0	128
C4	0	0	0	402	0	0	0	0	402
C5	0	0	0	0	102	0	0	0	102
C6	0	0	0	0	0	293	3	0	296
C7	0	0	0	0	0	5	148	0	153
C8	0	0	0	0	0	0	0	107	107
Total	167	285	128	402	102	298	151	107	1,640

(b) Multi-Layer Perceptron

Ground Truth Categories	Classifier's Categories								
	C1	C2	C3	C4	C5	C6	C7	C8	Total
C1	157	10	0	0	0	0	0	0	167
C2	1	282	0	0	2	0	0	0	285
C3	0	0	128	0	0	0	0	0	128
C4	4	0	0	389	9	0	0	0	402
C5	0	0	2	2	98	0	0	0	102
C6	0	0	1	0	0	260	25	10	296
C7	0	0	0	0	0	60	93	0	153
C8	0	0	0	0	0	3	0	104	107
Total	162	292	131	391	109	323	118	114	1,640

(c) Gaussian Maximum Likelihood

Ground Truth Categories	Classifier's Categories								
	C1	C2	C3	C4	C5	C6	C7	C8	Total
C1	161	5	0	1	0	0	0	0	167
C2	0	284	0	0	1	0	0	0	285
C3	0	0	124	0	4	0	0	0	128
C4	0	4	0	385	13	0	0	0	402
C5	0	0	0	0	102	0	0	0	102
C6	0	0	3	0	0	214	62	17	296
C7	0	0	0	0	0	37	116	0	153
C8	0	0	0	0	0	3	0	104	107
Total	161	293	127	386	120	254	178	121	1,640

Table A.2 Out-of-sample performance: Classification error matrices

(a) Fuzzy ARTMAP

Ground Truth Categories	Classifier's Categories								
	C1	C2	C3	C4	C5	C6	C7	C8	Total
C1	83	0	0	0	0	0	0	0	83
C2	0	142	0	0	0	0	0	0	142
C3	0	0	64	0	0	0	0	0	64
C4	0	0	0	200	0	0	0	0	200
C5	0	0	0	0	52	0	0	0	52
C6	0	0	0	0	0	146	2	0	148
C7	0	0	0	0	0	2	75	0	77
C8	0	0	0	0	0	0	0	54	54
Total	83	142	64	200	52	148	77	54	820

(b) Multi-Layer Perceptron

Ground Truth Categories	Classifier's Categories								
	C1	C2	C3	C4	C5	C6	C7	C8	Total
C1	79	4	0	0	0	0	0	0	83
C2	1	134	6	0	1	0	0	0	142
C3	0	0	64	0	0	0	0	0	64
C4	3	2	0	194	1	0	0	0	200
C5	0	3	0	0	49	0	0	0	52
C6	0	0	0	0	0	115	30	3	148
C7	0	0	0	0	0	29	48	0	77
C8	0	0	0	0	0	1	0	53	54
Total	83	143	70	194	51	145	78	56	820

(c) Gaussian Maximum Likelihood

Ground Truth Categories	Classifier's Categories								
	C1	C2	C3	C4	C5	C6	C7	C8	Total
C1	80	3	0	0	0	0	0	0	83
C2	0	141	0	0	1	0	0	0	142
C3	0	0	62	0	1	1	0	0	64
C4	1	3	0	191	5	0	0	0	200
C5	0	5	0	0	47	0	0	0	52
C6	0	0	1	0	2	73	64	8	148
C7	0	0	0	0	0	24	53	0	77
C8	0	0	0	0	0	2	0	52	54
Total	81	152	63	191	56	100	117	60	820

Appendix B
In-Sample and Out-of-Sample Map User's and Map Producer's Accuracies of the Classifiers

Table B.1 In-sample map user's and map producer's accuracies

Category	Name	Map User's Accuracy			Map Producer's Accuracy		
		Fuzzy ARTMAP	Multi-Layer Perceptron	Gaussian Maximum Likelihood	Fuzzy ARTMAP	Multi-Layer Perceptron	Gaussian Maximum Likelihood
C1	Mixed grass & arable farmland	100.0	94.0	96.4	100.0	96.9	95.1
C2	Vineyards & areas with low vegetation cover	100.0	98.9	99.6	100.0	96.9	96.9
C3	Asphalt & concrete surfaces	100.0	100.0	96.9	100.0	97.7	97.7
C4	Woodlands & public gardens with trees	100.0	96.8	95.8	100.0	99.5	97.7
C5	Low density residential & industrial areas [suburban]	100.0	96.1	100.0	100.0	89.9	87.3
C6	Densely built up residential & industrial areas [urban]	99.0	87.8	72.3	98.3	80.5	79.8
C7	Water courses	96.7	60.8	75.8	98.0	78.8	78.8
C8	Stagnant water bodies	100.0	97.2	97.2	100.0	91.2	85.8

Notes: Map user's accuracies for land use categories are calculated by dividing the number of correctly classified pixels in each category [i.e. the main diagonal elements of the classification error matrix] by the row totals.

Map producer's accuracies for land use categories are calculated by dividing the numbers of correctly classified pixels in each category [i.e. the main diagonal elements of the classification error matrix] by the columns totals.

Table B.2 Out-of-sample map user's and map producer's accuracies

Category	Name	Map User's Accuracy			Map Producer's Accuracy		
		Fuzzy ARTMAP	Multi-Layer Perceptron	Gaussian Maximum Likelihood	Fuzzy ARTMAP	Multi-Layer Perceptron	Gaussian Maximum Likelihood
C1	Mixed grass & arable farmland	100.0	95.2	96.4	100.0	95.2	98.8
C2	Vineyards & low vegetation cover areas	100.0	94.4	99.3	100.0	93.7	92.8
C3	Asphalt & concrete surfaces	100.0	100.0	96.9	100.0	91.4	98.4
C4	Woodlands & public gardens with trees	100.0	97.0	95.5	100.0	100.0	100.0
C5	Low density residential & industrial areas [suburban]	100.0	94.2	90.4	100.0	96.1	83.9
C6	Densely built up residential & industrial areas [urban]	98.6	77.7	49.3	98.6	79.3	73.0
C7	Water courses	97.4	62.3	68.8	97.4	61.5	45.3
C8	Stagnant water bodies	100.0	98.1	96.3	100.0	86.9	86.7

Notes: Map user's accuracies for land use categories are calculated by dividing the number of correctly classified pixels in each category [i.e. the main diagonal elements of the classification error matrix] by the row totals.

Map producer's accuracies for land use categories are calculated by dividing the number of correctly classified pixels in each category [i.e. the main diagonal elements of the classification error matrix] by the column totals.

8 Neural Spatial Interaction Models

Manfred M. Fischer
Department of Economic Geography & Geoinformatics,
Vienna University of Economics and Bu ness Administration

8.1 Introduction

One of the major intellectual achievements and, at the same time, perhaps the most useful contribution of spatial analysis to social science literature has been the development of spatial interaction models. Spatial interaction can be broadly defined as movement of people, commodities, capital and/or information over geographical space (see Batten and Boyce 1986). Such interaction encompasses such diverse behaviour as migration, travel-to-work, shopping, recreation, commodity flows, capital flows, communication flows [e.g. telephone calls], airline passenger traffic, the choice of health care services, and even the attendance at events such as conferences, cultural and sport events (Haynes and Fotheringham 1984). In each case, it results from a decision process in which an individual trades off in some way the benefit of the interaction with the costs entailed in overcoming the spatial separation between the point of departure and the destination. It is the pervasiveness of this type of trade-off in spatial behaviour which has made spatial interaction analysis and modelling so important, and the subject of intensive investigation in human geography and regional science (Fotheringham and O'Kelly 1989).

Since the pioneering work of Wilson (1970) on entropy maximization, however, there has been surprisingly little innovation in the design of spatial interaction models. The principal exceptions include the competing destinations version of Fotheringham (1983), the use of genetic algorithms to breed new forms of spatial interaction models, either directly (Openshaw 1988) or by genetic programming (Turton, Openshaw and Diplock 1997), and the design of neural spatial interaction models (Fischer and Gopal 1994a, b; Gopal and Fischer 1993; Openshaw 1993). Neural spatial models are termed 'neural' since they are based on neural computational approaches inspired by neuroscience. They are more closely related to spatial interaction models of the gravity type, and under commonly met conditions they can be viewed as a special class of general feedforward neural network models with a single hidden layer and sigmoidal transfer functions (see

Fischer 1998b). Rigorous mathematical proofs for the universality of such feedforward neural network models (see, among others, Hornik, Stinchcombe and White 1989) have established neural spatial interaction models as a powerful class of universal approximators for spatial interaction flow data.

Learning from examples, the problem for which neural networks were designed for to solve, is one of the most important research topics in artificial intelligence. A possible way to formalize learning from examples is to assume the existence of a function representing the set of examples, thus making it possible to generalize. This can be called a function reconstruction from sparse data [or, in mathematical terms, depending on the level of precision required, approximation or interpolation]. Within this general framework, the issues of key interest are the representational power of a given network model [in other words, the problem of model selection] and the procedures for obtaining optimal network parameters (see Fischer, Hlavácková-Schindler and Reismann 1999). The tasks of parameter estimation and model selection [or, in other words the specification of a network topology] are of crucial importance for the success of real world neural network applications.

The chapter is organized as follows. First, in the next section a brief description of single hidden layer neural spatial interaction models is given. Section 8.3 then focuses on the parameter estimation issue, pointing out that current practice is dominated by gradient-based local minimization techniques. These find local minima efficiently and work well in unimodal minimization problems, but can get trapped in multimodal problems. An alternative optimization scheme is provided by global search procedures. Section 8.4 describes the advantages of differential evolution, as an efficient direct search procedure for optimizing real-valued multi-modal objective functions. The specification of an adequate network topology is a key issue and the primary focus of Section 8.5. The problem of evaluating a neural spatial interaction model with a bootstrapping procedure is then addressed in Section 8.6. The proposed approach combines the purity of splitting the data into three disjoint sets – as suggested in Fischer and Gopal (1994a, b) – with the power of a resampling procedure. This provides a better statistical picture of forecast variability, and has the ability to estimate the effect of randomness of data splits versus randomness of the initial conditions of model parameters. The final section presents some conclusions and an outlook for the future.

8.2 The Model Class under Consideration

Suppose we are interested in approximating a N-dimensional spatial interaction function $F: R^N \rightarrow R$, where R^N as N-dimensional Euclidean real space is the input space and $\mathfrak{R}$ as 1-dimensional Euclidean real space is the output space. This function should estimate spatial interaction flows from regions of origin to regions of destination. The function F is not explicitly known, but given by a finite set of samples $S=\{(x_k, y_k), k=1, ..., K\}$ so that $F(x_k)=y_k$, $k=1, ..., K$. The set S is the set of pairs of input and output vectors. The task is to find a continuous function that approximates set S. In real world applications, K is a small number and the samples contain noise.

To approximate F, we consider the class of neural spatial interaction models Ω with one hidden layer, N input units, J hidden units and one output unit. Ω consists of a composition of transfer functions so that the [single] output of Ω is

$$y = \Omega(\boldsymbol{x}, \boldsymbol{w}) = \psi\left(\sum_{j=0}^{J} w_j \, \phi_j\left(\sum_{n=0}^{N} w_{jn} \, x_n\right)\right). \tag{8.1}$$

The expression $\Omega(\boldsymbol{x}, \boldsymbol{w})$ is a convenient short-hand notation for the model output since this depends only on inputs and weights. Vector $\boldsymbol{x} = (x_1, ..., x_N)$ is the input vector augmented with a bias signal x_0, which can be thought of as being generated by a 'dummy unit' whose output is clamped at 1. The w_{jn}'s represent input to hidden connection weights and the w_j's hidden to output weights [including the bias]. The symbol $\boldsymbol{w}$ is a shorthand notation of the d=$(J(N+1)+J+1)$-dimensional vector of all the w_{jn} and w_j network weights and biases [i.e. model parameters]. $\phi_j(.)$ and $\psi(.)$ are differentiable non-linear transfer functions of the hidden units $j = 1, ..., J$ and the output unit, respectively.

The architecture of the single hidden layer feedforward network model determines the precise form of the function Ω. Different network architectures will lead to different forms for Ω. Each neural spatial interaction model $\Omega(\boldsymbol{x}, \boldsymbol{w})$ can be represented in terms of a network diagram (see Fig. 8.1) such that there is a one-to-one correspondence between components of Ω and the elements of the diagram. Equally, any topology of a three layer network diagram with three inputs and a single output, provided it is feedforward, can be translated into the corresponding neural spatial interaction model with N=3. We can thus consider model choice in terms of topology selection [i.e. choice of the number of hidden units] and specification of the transfer functions ψ and ϕ_j (j=1, ..., J).

Different types of transfer functions ϕ_j (j=1, ..., J) and ψ will lead to different particular computational networks. If the transfer functions are taken to be linear, functional form (8.1) becomes a special case of the general linear regression model. The crucial difference is that here we consider the weight parameters appearing in the hidden and output layers as being adaptive, so their values can be changed during the process of network training [in statistical terminology: parameter estimation].

The novelty and fundamental contribution of the feedforward neural network approach to spatial interaction analysis derive not so much on the associated learning methods, which will be discussed in Section 8.4, as its focus on functions such as ϕ_j and ψ in (8.1). Equation (8.1) is a rather general class of spatial interaction functions, and one which took a long time to be appreciated. It has been shown by Hornik, Stinchcombe and White (1989) and other authors that neural networks with a single hidden layer, such as (8.1), under generally met conditions can approximate any continuous function Ω uniformly on compacta, by increasing the size of J of the hidden layer. Because of this universal approximation property, single hidden layer feedforward network models are useful for applications in spatial analysis in general and spatial interaction analysis in particular, and a host of related tasks. There are also some results on the rate of approximation [i.e. how many hidden units are needed to approximate to a specified accuracy], but as always with such results, they are no guide to how many hidden units might be needed in any practical problem.

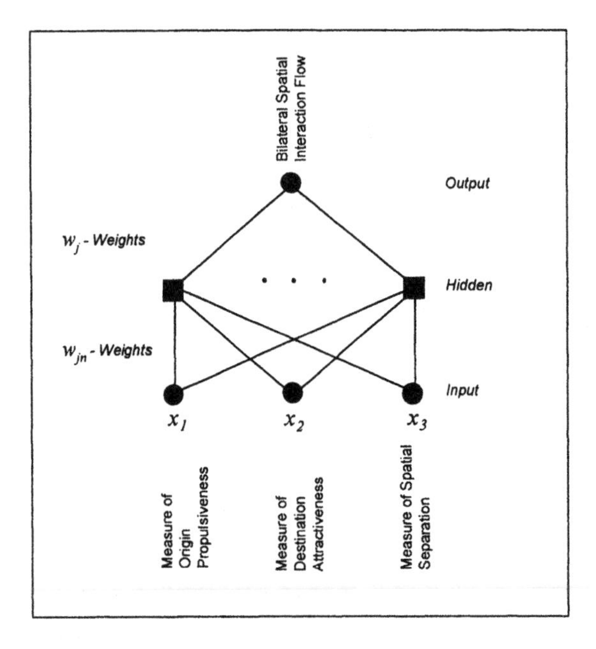

Fig. 8.1 Representation of the general class of neural spatial interaction models defined by Equation (8.1) for N=3 [biases not shown]

Neural spatial interaction modelling may be viewed as a *three-stage process* as outlined in Fischer and Gopal (1994a):

- The *first stage* consists of the identification of a model candidate from the class (8.1) of neural spatial interaction models [i.e. the model choice task]. This involves both the specification of appropriate transfer functions ψ and ϕ_j (j = 1, ..., J) and the determination of an adequate network topology of Ω.
- The *second stage* involves the network training problem [parameter estimation problem], that is the determination of an optimal set of model parameters where optimality is defined in terms of an error or performance function.

- The *third stage* is concerned with testing and evaluating the out-of-sample [generalization] performance of the chosen model.

As already mentioned, the model selection and parameter estimation are of crucial importance to real world neural spatial interaction modelling. Obviously, these two processes are intertwined. If an appropriate J and a good set of transfer functions can be found, the success of which depends on the particular real world problem, then the task of weight learning [parameter estimation] generally becomes easier to perform.

Following Fischer and Gopal (1994a, b), it seems useful to consider the case of N=3, i.e. the input space will be a closed interval of the three-dimensional Euclidean space R^3. The three input units are chosen to correspond to the independent variables of the classical unconstrained spatial interaction model. They represent measures of origin propulsiveness, destination attractiveness and spatial separation. The output unit corresponds to the dependent variable of the classical model and represents the spatial interaction flows from origin to destination. The above general framework for neural spatial interaction analysis has been used for many years. Experience has found the logistic function to be an appropriate choice for ψ and ϕ_j (j = 1, ..., J). This form of specification of the transfer functions leads to the following special class $\Omega_L(\boldsymbol{x}, \boldsymbol{w})$ of spatial interaction functions $\Omega(\boldsymbol{x}, \boldsymbol{w})$

$$\Omega_L(\boldsymbol{x},\boldsymbol{w}) = \left\{1+\exp\left[-\sum_{j=0}^{J} w_j \left(1+\exp\left(-\sum_{n=0}^{N} w_{jn}\, x_n\right)\right)^{-1}\right]\right\}^{-1} \tag{8.2}$$

with N=3. Such models may be viewed as non-linear regression models of a quite specific form. If is ψ taken to be the identity function we arrive at

$$\Omega_L^I(\boldsymbol{x},\boldsymbol{w}) = \sum_{j=0}^{J} w_j \left(1+\exp\left(-\sum_{n=0}^{N} w_{jn}\, x_n\right)\right)^{-1} \tag{8.3}$$

A heuristic reason why neural spatial interaction models of type (8.2) or (8.3) might work well with modest numbers of hidden units is that the first stage allows a projection onto a subspace of R^N within which the approximation can be performed. In this feature, these neural models share many of the properties of projection pursuit regression. Indeed, for theoretical purposes (8.3) is essentially equivalent to projection pursuit regression (see Friedman and Stuetzle 1981).

Since the above neural network models can perform essentially arbitrary non-linear functional mappings, a neural spatial interaction model could – in principle – be used to map the raw input data directly onto the required final output values.

In practice, for all but the simplest problems, such an approach will generally give poor results for a number of reasons. For most applications, it is necessary first to transform the data into some new representation before training a model. In the simplest case, pre-processing may take the form of a linear transformation of the input data, and possibly also of the output data [post-processing]. More complex pre-processing may involve reduction of the dimensionality of the input data.

Another important way in which model performance can be improved is through the incorporation of *prior knowledge*. This refers to relevant information [such as production or attraction constraints in the spatial interaction system] that might be used to develop a solution, and is additional to information provided by the training data. Prior knowledge can either be incorporated into the network structure itself or into the pre-processing and post-processing stages. It can also be used to modify the training process through the use of regularization, as discussed in Section 8.5.

8.3 Training Neural Spatial Interaction Models: Classical Techniques

In this section we restrict our scope to the network training problem and to training algorithms that train a fixed member of the class Ω of spatial interaction models. The approximation of Ω to F then only depends on the learning samples, and the learning [training] algorithm that determines the parameter $\boldsymbol{w}$ from S and the model specification.

The Parameter Estimation Problem

Let us assume the hidden transfer function $\phi_j(.)$ to be identical, $\phi_j(.)=\psi(.)$ for all $j = 1, ..., J$, and the logistic function. Moreover, the model has a fixed topology, i.e. N=3 and J are predetermined. The goal of learning is to find suitable values $\boldsymbol{w}^*$ for the network weights of the model such that the underlying mapping $F: R^3 \rightarrow R$ represented by the training set of samples is approximated or learned. The process of determining optimal parameter values is called training, or learning, and can be formulated in terms of the minimization of an appropriate error function [or cost function] E to measure the degree of approximation with respect to the actual setting of network weights. The most common error function is the square-error function of the patterns over the finite set of training data, so that the parameter estimation problem may be defined for batch learning as the following minimization problem:

$$\min_{\boldsymbol{w}} E(\boldsymbol{w}, S) = \min_{\boldsymbol{w}} \sum_{(x_k, y_k) \in S} \left| y_k - \Omega\,(x_k, \boldsymbol{w}) \right|^2 \tag{8.4}$$

where the minimization parameter is the weight vector, w, defining the search space. The function $E(w, S)$ is non-negative continuously differentiable on the d-dimensional parameter space, which is a finite dimensional closed bounded domain, and thus compact. So $E(w, S)$ assumes its minimum value w^* as the weight minimum. Characteristically, there exist many minima all of which satisfy

$$\nabla E(w, S) = 0 \tag{8.5}$$

where $\nabla E(w, S)$ denotes the gradient error function in the d-dimensional parameter space. The minimum for which the value of $E(w, S)$ is smallest is termed the global minimum, while other minima are called local minima. But there is no guarantee about what kind of minimum is encountered. Due to the non-linearity of the transfer functions it is not possible to find closed-form solutions for this optimization problem.

There are two basic approaches to finding the minimum of the global error function E – off-line learning and on-line learning. These differ in how often the weights are updated. The *on-line learning* [that is, pattern based learning] updates the weights after every single pattern s_k chosen at random from S, that is, using only information from one pattern. On the other hand, *off-line learning* updates the weights after K' patterns randomly chosen from S have been propagated through the network, that is, using information from K' patterns in the training set. If $K'=K$, off-line learning is known as *batch learning*, otherwise it is called *epoch-based learning* with an epoch size of K' ($1<K'<K$). Neither on-line or epoch-based (K' small) versions are consistent with optimization theory, but have nevertheless been found superior to batch learning on real world problems with a realistic level of complexity and have a training set that goes beyond a critical threshold (see Le Cun 1989; Schiffmann, Jost and Werner 1993; Fischer and Staufer 1999, for the spectral pattern classification problem).

The Optimization Strategy

Above the network training problem has been formulated as a problem of minimizing the least square error function and, thus, a special case of function approximation where no explicit model of the data is assumed. Most of the optimization procedures used to minimize functions are based on the same strategy. The minimization is a local iterative process in which an approximation to the function in a neighbourhood of the current point in parameter space is minimized. The approximation is often given by a first- or second-order Taylor expansion of the function. In the case of batch learning, the general scheme of the iteration process may be formulated as follows:

(i) choose an initial vector w in parameter space and set $\tau = 1$

(ii) determine a search direction $d(\tau)$ and a step size $\eta(\tau)$ so that

$$E(w(\tau) + \eta(\tau)\, d(\tau)) < E(w(\tau)) \qquad \tau = 1, 2, \ldots \tag{8.6}$$

(iii) update the parameter vector

$$w(\tau+1) = w(\tau) + \eta(\tau)\, d(\tau) \qquad \tau = 1, 2, \ldots \tag{8.7}$$

(iv) if $dE(w)/dw \neq 0$ then set $\tau = \tau + 1$ and go to (ii), else return $w(\tau + 1)$ as the desired minimum.

In the case of *on-line learning*, the above scheme has to be slightly modified since this learning approach is based on the [local] error function E_k, and the parameter vector $w_k(\tau)$ is updated after every presentation of $s_k = (x_k, y_k)$. In both cases, batch and on-line learning, determining the next current point in the iteration process entails two problems. *First*, the search direction $d(\tau)$ has to be determined, that is, in what direction in parameter space we want to go in the search for a new current point. *Second*, once the search direction has been found, we have to decide how far to go in the specified direction, that is, a step size $\eta(\tau)$ has to be determined. To solve these problems, normally two types of operation must be carried out: the computation or the evaluation of the derivatives of the error function with respect to the network parameters, and the computation of the parameter $\eta(\tau)$ and the direction vector $d(\tau)$ based upon these derivatives. The evaluation of the derivatives of the error function is most commonly performed by the backpropagation technique which provides a computationally efficient procedure for doing this (Rumelhart, Hinton and Williams 1986).

Optimization Techniques for Parameter Adjustments

In numerical optimization, there are various techniques for the computation of the parameter $\eta(\tau)$ and the direction vector $d(\tau)$ (see, for example, Luenberger 1984; Fletcher 1986). In particular, three techniques have gained popularity:

(i) The *steepest-descent* [gradient] [GD] *methods* calculate the actual search direction $d(\tau)$ as the negative gradient

$$d(\tau) := -\nabla E(w(\tau)) \qquad \tau = 1, 2, \ldots \tag{8.8}$$

(ii) *Conjugate gradient* [CG] *methods* calculate the actual search direction $d(\tau)$ as a linear combination of the gradient vector and the previous search directions (see Hestenes and Stiefel 1952). In the PR-CG algorithm, the Polak-Ribiere variant of conjugate gradient procedures (see Press et al. 1992), the search direction is computed as

$$d(\tau) := -\nabla E(w(\tau)) + \beta(\tau)\, d(\tau-1) \qquad \tau = 1, 2, \ldots \tag{8.9}$$

with

$$d(0) = -\nabla E(w(0)) \tag{8.10}$$

where $\beta(\tau)$ is a scalar parameter that ensures that the sequence of vectors $\boldsymbol{d}(\tau)$ satisfying the following condition expressed as

$$\beta(\tau) = \frac{[\nabla E(\boldsymbol{w}(\tau)) - \nabla E(\boldsymbol{w}(\tau-1))]^T \, \nabla E(\boldsymbol{w}(\tau))}{\nabla E(\boldsymbol{w}(\tau-1))^T \, \nabla E(\boldsymbol{w}(\tau-1))} \tag{8.11}$$

$\boldsymbol{w}(\tau-1)^T$ is the transpose of $\boldsymbol{w}(\tau-1)$. Note that the CG algorithm utilizes information about the direction search $\boldsymbol{d}(\tau-1)$ from the previous iteration in order to accelerate convergence, and each search direction would be conjugate if the objective function would be quadratic.

(iii) *Quasi-Newton* – or variable metric – *methods* employ the differences of two successive iteration points τ and $\tau + 1$, and the difference of the corresponding gradients to approximate the inverse Hessian matrix. The most commonly used update technique is the BFGS [Broyden-Fletcher-Goldfarb-Shanno] algorithm (see Luenberger 1984; Fletcher 1986; Press et al. 1992) that determines the search direction as

$$\boldsymbol{d}(\tau) = -\boldsymbol{H}(\tau)\nabla E(\boldsymbol{w}(\tau)) \tag{8.12}$$

where $\boldsymbol{H}(\tau)$ is some $\boldsymbol{w}\text{x}\boldsymbol{w}$ symmetric positive definite matrix and denotes the current approximation to the inverse of the Hessian matrix, i.e.,

$$\boldsymbol{H}(\tau) \equiv [\nabla^2 E(\boldsymbol{w}(\tau))]^{-1} \tag{8.13}$$

where

$$\boldsymbol{H}(\tau) = \left\{ \mathrm{I} - \frac{\boldsymbol{d}(\tau-1)\,(\boldsymbol{g}(\tau-1))^T}{(\boldsymbol{d}(\tau-1))^T \, \boldsymbol{g}(\tau-1)} \right\} \boldsymbol{H}(\tau-1) \left\{ \mathrm{I} - \frac{\boldsymbol{g}(\tau-1)\,(\boldsymbol{d}(\tau-1))^T}{(\boldsymbol{d}(\tau-1))^T \, \boldsymbol{g}(\tau-1)} \right\} + \frac{\boldsymbol{d}(\tau-1)\,(\boldsymbol{d}(\tau-1))^T}{\boldsymbol{d}(\tau-1)^T \, \boldsymbol{d}(\tau-1)} \tag{8.14}$$

with

$$\boldsymbol{g}(\tau-1) := -\nabla E(\boldsymbol{w}(\tau)) - \nabla E(\boldsymbol{w}(\tau-1)). \tag{8.15}$$

$\boldsymbol{H}$ is initialized usually with the identity matrix $\boldsymbol{I}$ and updated at each iteration using only gradient differences to approximate second-order information. The inverse Hessian is more closely approximated as iterations proceed.

Both the PR-CG and the BFGS algorithms raise the calculation complexity per training iteration considerably, since they have to perform a one-dimensional linear search in order to determine an appropriate step size. A line search involves several calculations of the error function E or its derivative, both of which increase the complexity. Characteristically, the parameter $\eta = \eta(\tau)$ is chosen to minimize

$$E(\tau) = E(\boldsymbol{w}(\tau)) + \eta\, \boldsymbol{d}(\tau) \tag{8.16}$$

in the τ-th iteration. This gives an automatic procedure for setting the step length, once the search direction $\boldsymbol{d}(\tau)$ has been determined.

All three procedures use only first-order derivative information of the error function. The derivatives can thus be calculated efficiently by backpropagation as shown in Fischer and Staufer (1999). The steepest descent algorithm has the great advantage of being very simple and cheap to implement. One of its limitations is the need to choose a suitable step size η by trial and error. Inefficiency is primarily due to the risk that the minimization directions and step sizes may be poorly chosen. Unless the first step is chosen such that it leads directly to the minimum, steepest descent will zigzag with many small steps. It is worth noting that there have been numerous attempts in recent years to improve the performance of the basic gradient descent by making various ad hoc modifications (see, for example, Jacobs 1988), such as the heuristic scheme known as *quickprop* (Fahlmann 1988).

In contrast, the conjugate gradient and quasi-Newton procedures are intrinsically off-line parameter adjustment techniques, and evidently more sophisticated optimization procedures. In terms of complexity and convergence properties, the conjugate gradient can be regarded as falling in between the gradient descent method and the quasi-Newton technique (Cichocki and Unbehauen 1993). Its advantage is the simplicity for estimating optimal values of the coefficients $\eta(\tau)$ and $\beta(\tau)$ at each iteration. No $\boldsymbol{w}$x$\boldsymbol{w}$-dimensional matrices $\boldsymbol{H}(\tau)$ need to be generated as in the quasi-Newton procedures. The search direction is chosen by setting the β appropriately, so that $\boldsymbol{d}$ distorts as little as possible the minimization achieved by the previous search step. A major difficulty is that, for a non-quadratic error function, the directions obtained are *not* necessarily descent directions and numerical instability can result (Battiti and Tecchiolli 1994). Periodically, it might be necessary to restart the optimization process by a search in the steepest descent direction when a non-descent search direction is generated. It is worth mentioning that the gradient descent procedures can be viewed as a form of gradient descent with an adaptive momentum $\beta(\tau)$, the important difference being that $\eta(\tau)$ and $\beta(\tau)$ in conjugate gradient are automatically determined at each iteration (see Equations (8.11) and (8.16)).

The conjugate gradient methods are not as effective as some quasi-Newton procedures. It is noteworthy that quasi-Newton techniques are a shortcut for the Newton-Raphson algorithm to speed up computations when the derivative calculation is time consuming. They require approximately twice as many gradient evaluations as the quasi-Newton methods. However, they save time and memory required for calculating the $\boldsymbol{w}$x$\boldsymbol{w}$-dimensional matrices $\boldsymbol{H}(\tau)$, especially in the case of large-sized problems (Shanno 1990). In quasi-Newton procedures, the matrices $\boldsymbol{H}(\tau)$ are positive definite approximations of the inverse Hessian matrix obtained from gradient information. Thus, it is not necessary to evaluate second-order derivatives of E. A significant advantage of the quasi-Newton over the conjugate gradient procedures is that line search need not be performed with such accuracy, since it is not a critical feature in the algorithms. Scaled conjugate gradient algorithms provide a means of avoiding time-consuming line search of conventional conjugate gradients by utilizing the model-trust region approach, from the Levenberg-Marquardt algorithm, a variation of the standard Newton algorithm (see Møller 1993 for more details). Algorithms, such as BFGS, are always stable since $\boldsymbol{d}(\tau)$ is always a descent search direction. Today, they are the most efficient and sophisticated optimization techniques for batch training, but are expensive both in terms of computation and memory. Large-sized real world problems implying greater $\boldsymbol{w}$ could lead to prohibitive memory requirements (Shanno 1990).

In general, the above minimization techniques find the local minima efficiently and work best in unimodal problems. They have difficulties when the surface of the parameter space is flat [i.e. gradients close to zero], when there is a large range of gradients, and when the surface is very rugged. The search may progress too slowly when the gradient is small, and may overshoot where the gradient is large. When the error surface is rugged, a local search from a random starting point generally converges to a local minimum close to the initial point and a worse solution than the global minimum.

8.4 A New Global Search Approach for Network Training: The Differential Evolution Method

Global search algorithms employ heuristics to be able to escape from local minima. These algorithms can be classified as probabilistic or deterministic. Of the few deterministic global minimization methods developed, most apply deterministic heuristics to bring search out of a local minimum. Other methods, like covering methods, recursively partition the search space into subspaces before searching. None of these methods operates well or provides adequate coverage when the search space is large, as is usually the case in neural spatial interaction modelling.

Probabilistic global minimization methods rely on probability to generate decisions. The simplest probabilistic algorithm uses restarts to bring search out of a local minimum when little improvement can be made locally. More advanced methods rely on probability to indicate whether a search should ascend from a local minimum: simulated annealing, for example, when it accepts uphill movements. Other probabilistic algorithms rely on probability to decide which intermediate points to interpolate as new trial parameter vectors: random recombinations or mutations in evolutionary algorithms (see, for example, Fischer and Leung 1998).

Central to global search procedures is a strategy that generates variations in the parameter vectors. Once a variation is generated, a decision has to be made whether or not to accept the newly derived trial parameter. Standard direct search methods [with few exceptions such as simulated annealing] utilize the greedy criterion to make the decision. Under this criterion, a new parameter vector is accepted if and only if it reduces the value of the error function. Although this decision process converges relatively fast, it carries the risk of entrapment in a local minimum. Some stochastic search algorithms, like genetic algorithms and evolution strategies, employ a multipoint search strategy to escape from local minima.

The Differential Evolution Method [DEM], originally developed by Storn and Price (1996, 1997) and adopted by Fischer, Hlavácková-Schindler and Reismann (1999) for spatial interaction analysis, is a global optimization algorithm that employs a structured, yet randomized, parallel multipoint search strategy which is biased towards reinforcing search points at which the error function $E(\boldsymbol{w}, S)$ has relatively low values. The DEM is similar to simulated annealing in that it employs a random [probabilistic] strategy. But one of the distinguishing features of DEM is its effective implementation of parallel multipoint search. DEM maintains a collection of samples from the search space rather than a single point. This collection of samples is called a *population* of trial solutions.

To start the stochastic multipoint search, an initial population P of, say M, d-dimensional parameter vectors $P(0) = \{w_0(0), ..., w_{M-1}(0)\}$ is created. For the parameter estimation problem at hand, $d = J(N+1)+J+1$. Usually this initial population is created randomly, because it is not known a priori where the globally optimal parameter is likely to be found in the parameter space. If such information is given, it may be used to bias the initial population towards the most promising regions of the search space by adding normally distributed random deviations to this solution candidate given a priori. From this initial population, subsequent populations $P(1)$, $P(2)$, ..., $P(\tau)$, ... will be computed by a scheme that generates new parameter vectors by adding the weighted difference of two vectors to a third. If the resulting vector yields a lower error function value than a predetermined population member, the newly generated vector will replace the vector it was compared to, otherwise the old vector is retained. As in evolution strategies, the greedy criterion is used in the iteration process, and the probability distribution functions determining vector mutations are not given a priori. The scheme for generating $P(\tau + 1)$ from $P(\tau)$ with $\tau \geq 0$ can be summarized by three major stages: the construction of a $\boldsymbol{v}(\tau+1)$-vector from vector $\boldsymbol{w}(\tau)$-vectors [Stage 1], the construction of a $\boldsymbol{u}(\tau+1)$-vector from $\boldsymbol{v}(\tau+1)$ and $\boldsymbol{w}(\tau)$-vectors [Stage 2], and the decision criterion whether or not the $\boldsymbol{u}(\tau+1)$-vector should become member of the population [Stage 3] representing a possible solution of the parameter estimation problem under consideration at step $\tau+1$ of the iteration process. The iteration process continues until some stopping criterion applies.

Stage 1: For each population member $w_m(\tau)$, $m = 0, 1, ..., M-1$, a perturbed vector $v_m(\tau+1)$ is generated according to:

$$v_m(\tau+1) = w_{best}(\tau) + \kappa(w_{r_1}(\tau) - w_{r_2}(\tau)) \tag{8.17}$$

with r_1, r_2 integers chosen randomly from $\{0, ..., M-1\}$ and mutually different. The integers are also different from the running index m. $\kappa \in (0, 2]$ is a real constant factor which controls the amplification of the differential variation $(w_{r_1}(\tau) - w_{r_2}(\tau))$. The parameter vector $w_{best}(\tau)$ which is perturbed to yield $v_m(\tau+1)$ is the best parameter vector of population $P(\tau)$.

Stage 2: In order to increase the diversity of the new parameter vectors, some specific type of crossover may be introduced. One might use the crossover of the exponential type (Storn and Price 1996), yielding the vector:

$$u_m(\tau+1) = (u_{0m}(\tau+1),\ldots, u_{(d-1)m}(\tau+1)) \tag{8.18}$$

where

$$u_{km}(\tau+1) = \begin{cases} v_{km}(\tau+1) & \text{for } k = \langle i \rangle_d, \ldots, \langle i+R-1 \rangle_d \\ w_{km}(\tau) & \text{for all other } k \in [0, d-1] \end{cases} \tag{8.19}$$

is formed. The brackets $\langle\,\rangle_d$ denote the modulo function with modulus d. In other words, a sequence of R coordinates of vector $\boldsymbol{u}(\tau+1)$ is identical to the corresponding coordinates of vector $\boldsymbol{v}(\tau+1)$, whereas the other coordinates of $\boldsymbol{u}(\tau+1)$ are retained as the original values of $\boldsymbol{w}(\tau)$. The starting index i in (8.19) is a randomly chosen integer from the interval $[0, d-1]$. The integer R, which denotes the number of parameters that are going to be exchanged, is drawn from the interval $[1, d]$ with the probability $Pr(R=\eta) = (\text{CR})^{\eta}$, where $\eta > 0$ and $CR \in [0, 1]$ is the crossover probability and forms a control variable for the scheme. The random decisions for both i and R are made anew for each newly generated trial vector $u_m(\tau+1)$. It is worth noting that $CR = 1$ implies $u_m(\tau+1) = v_m(\tau+1)$.

Stage 3: The decision whether or not $u_m(\tau+1)$ should become a member of $P(\tau+1)$, is based on the greedy criterion. If:

$$E(u_m(\tau+1)) < E(w_m(\tau)) \tag{8.20}$$

then $w_m(\tau)$ is replaced by $u_m(\tau+1)$ otherwise the old value $w_m(\tau)$ is retained as $w_m(\tau+1)$.

Using non-linear optimization algorithms such as DEM, the choice must be made when to stop the training process. Possible choices are listed below:

(i) Stop after a fixed number of iterations. The problem with this approach is that it is difficult to know a priori how many iterations would be appropriate. An approximate idea can be obtained from some preliminary tests.
(ii) Stop when the error function falls below some a pre-specified value. This criterion suffers from the problem that the specified value may never be reached, so a limit on iterations as in (i) is also required.
(iii) Stop when the relative change in error function falls below some pre-specified value. This may lead to premature termination if the error function decreases relatively slowly during some part of the training process.
(iv) Stop training when the error measured using an independent validation set starts to increase. This approach, called early stopping or cross-validation, may be used as part of a strategy to optimize the generalization performance of the network model (for details see Fischer and Gopal 1994a).

In practice, some combination of the above strategies may be employed as part of a largely empirical process of parameter estimation.

Taking Austrian interregional telecommunication traffic data as a testbed, Fischer, Hlavácková-Schindler and Reismann (1999) have found that the differential evolution method slightly outperforms backpropagation of conjugate gradients in terms of in-sample and out-of-sample performance, but at a very high computational costs. The issue of accuracy has ramifications with respect to the a priori knowledge of the response surface. If a correct neural spatial interaction model structure is assumed, DEM tends in general to be slower than conventional local optimization schemes, such as the conjugate gradient procedure used as benchmark. This results from the inefficiency of not using information about the gradient of the error function, although gradient methods could be incorporated in parallel with differential evolution. There is potential for further development of this latter strategy.

8.5 Selecting Neural Spatial Interaction Models: The Model Choice Issue

One of the central issues in neural network modelling in general, and neural spatial interaction modelling in particular, is that of model selection. The model specification problem includes, firstly, the specification of appropriate transfer functions ψ and ϕ_j (j=1, ..., J) and, secondly, the determination of an adequate network topology of Ω. In the case of model classes Ω_L and Ω_L^{I}, this means the determination of the optimum size of J. This is because an increase in J, in other words in the number of connection parameters, will generally allow a smaller error value to be found. The goal of model selection is to optimize the complexity of the model in order to achieve the best out-of-sample performance, that is, generalization (see Fischer 2000).

The Concept of the Bias-Variance Trade-Off

Considerable insight into this phenomenon can be achieved by introducing the concept of the bias-variance trade-off, in which the generalization error is decomposed into the sum of the bias squared plus the variance. Following Bishop (1995) the sum-of-squares error, in the limit of an infinite data set S, can be written in the form of

$$E(\boldsymbol{w}, S) = \frac{1}{2} \int (\Omega(\boldsymbol{x}, \boldsymbol{w}) - \langle y|\boldsymbol{x}\rangle)^2 \, \mathrm{p}(\boldsymbol{x}) \, \mathrm{d}\boldsymbol{x} + \frac{1}{2} \int (\langle y^2|\boldsymbol{x}\rangle - \langle y|\boldsymbol{x}\rangle^2) \, \mathrm{p}(\boldsymbol{x}) \, \mathrm{d}\boldsymbol{x} \qquad (8.21)$$

in which $\mathrm{p}(\boldsymbol{x})$ is the unconditional density of the input data, $\langle y|\boldsymbol{x}\rangle$ denotes the conditional average, or regression, of the target data y given by

$$\langle y|\boldsymbol{x}\rangle \equiv \int y \, \mathrm{p}(y|\boldsymbol{x}) \, \mathrm{d}y \qquad (8.22)$$

where $\mathrm{p}(y|\boldsymbol{x})$ is the conditional density of the target variable y conditioned on the output vector $\boldsymbol{x}$. Similarly

$$\langle y^2|\boldsymbol{x}\rangle \equiv \int y^2 \, \mathrm{p}(y|\boldsymbol{x}) \, \mathrm{d}y \qquad (8.23)$$

Note that the second term in (8.21) is independent of the spatial interaction function $\Omega(\boldsymbol{x}, \boldsymbol{w})$ and, thus, is independent of the network weights $\boldsymbol{w}$. The optimal model, in the sense of minimizing the sum-of-squares error, is the one that makes the first term in (8.21) vanish, and is given by $\Omega(\boldsymbol{x}, \boldsymbol{w}) = \langle y|\boldsymbol{x}\rangle$. The second term represents the intrinsic noise in the data and sets a lower limit on the error that can be achieved.

In real world applications, we have to deal with the problems arising from a finite size data set. Suppose, we consider a training set S consisting of K patterns that we utilize to determine the neural spatial interaction function $\Omega(\boldsymbol{x}, \boldsymbol{w})$. Now consider a whole ensemble of possible data sets, each containing K patterns, and each taken from the same fixed joint distribution $\mathrm{p}(\boldsymbol{x}, y)$. The optimal network model is given by the conditional average $\langle y|\boldsymbol{x}\rangle$. A measure of how close the actual spatial interaction function $\Omega(\boldsymbol{x}, \boldsymbol{w})$ is to the desired one is given by the integrand of the first term in (8.21): $(\Omega(\boldsymbol{x}, \boldsymbol{w})) - \langle y|\boldsymbol{x}\rangle)^2$. The value of this quantity will depend on the particular data set S on which it is trained. We can eliminate this dependence by considering an average over the complete ensemble of data sets, that is

$$\varepsilon_S [(\Omega(x, w) - \langle y|x\rangle)^2] \tag{8.24}$$

where $\varepsilon_S(.)$ denotes the expectation [ensemble average]. If model Ω were always a perfect predictor of the regression function $\langle y|x\rangle$ then this error would be zero. A non-zero error can arise essentially for two distinct reasons. First, it may be that the model Ω is on average different from the regression function. This is termed *bias*. Second, it may be that Ω is very sensitive to the particular data set S, so that for a given x, it is larger than the required value for some data sets, and smaller for other data sets. This is called *variance*. We can make the decomposition into bias and variance explicit by writing (8.24) in a somewhat different but mathematically equivalent form (see Bishop 1995):

$$\varepsilon_S [(\Omega (x, w) - \langle y|x\rangle)^2] = \varepsilon_S (\Omega (x, w) - \langle y|x\rangle)^2 + \varepsilon_S (\Omega (x, w) - \varepsilon_S (\Omega (x, w)))^2 \tag{8.25}$$

where the first term of the right hand side of the equation denotes the bias squared and the second term the variance. The bias measures the extent to which the average over all data sets of the spatial interaction function differs from the desired function $\langle y|x\rangle$. Conversely, the variance measures the extent to which Ω is sensitive to the particular choice of data sets.

A neural spatial interaction model that is too simple [i.e. with small J] or too inflexible will have a large bias and smooth out some of the underlying structure in the data [corresponding to high bias], while one that has too much flexibility in relation to the particular data set will overfit the data [corresponding to high variance] and have a large variance. In either case, the performance of the network on new data [i.e. generalization performance] will be poor. This highlights the need to optimize the complexity in the model selection process in order to achieve the best generalization.

Model Selection Techniques

Both the theoretical and practical side of the model selection problem has been intensively studied in the field of neural networks and a vast array of methods have been suggested to perform this task [see also Chapter 3 in this volume]. Most approaches view model selection as a process consisting of a series of steps that are performed independently.

Step 1: The first step consists of choosing a specific parametric representation [i.e. model] that is oversized in comparison to the size of the training set used.

Step 2: In this step either an error function such as E [possibly including a regularization term] is chosen directly or, in a Bayesian setting, prior distributions on the elements of the data generation process [noise, model parameter, regularisers, etc.] are specified, and an objective function is derived from these.

Step 3: Using the error function specified in *Step 2*, the training process is started and continued until a convergence criterion is fulfilled. The resulting parametrization of the given model architecture is then placed in a pool of model candidates from which the final model will be chosen.

Step 4: To avoid overfitting, model complexity must be limited. Thus, the next step usually consists of modifying the network model architecture [for example, by pruning weights] or the penalty term [for example, by changing its weighting in the objective function], or of the Bayesian prior distributions. The last two modifications lead to a modification of the objective function. It is worth noting that this establishes a new framework for the training process, which is then restarted and continued until convergence, yielding another model for the pool.

This process is iterated until the model builder is satisfied that the pool contains a reasonable diversity of model candidates. These are then compared with each other using some estimator of generalization ability, for example, the average relative variance [i.e. a normalized mean squared error metric] given by Fischer and Gopal (1994a).

$$\text{ARV}(S) = \frac{\sum_{(x_k, y_k) \in S} |y - \Omega(x_k, w)|^2}{\sum_{(x_k, y_k) \in S} (y_k - \bar{y})^2} \tag{8.26}$$

where y denotes the target vector and $\bar{y}$ the average over K desired values in S.

The methods employed for training can be very sophisticated (see, for example, Fischer and Staufer 1999). In contrast, the choice and modification of the network model architecture and objective function is generally ad hoc, or is directed by a search heuristic (see, for example, Openshaw 1993). In this section three principal approaches for directing model modification and selection are considered. The *first approach* uses *pruning techniques*. The principal idea of pruning is to reduce the number of model parameters by removing dispensable ones. Pruning techniques function by training an oversized neural network model with a fixed, but larger J to a minimum of $E(w, S)$, then testing elements of the model, such as connection parameters, for relevance. Those elements with poor test results are then deleted and the modified network model retrained. In this approach one uses the information in an existing model to direct the search to the best 'neighbouring' model.

Clearly, in utilizing this approach, various choices have to be made. The most important is how to decide which parameter weights should be removed. The decision is generally based on some measure of the relative importance, or saliency, of different weight parameters. The simplest concept of saliency is to suppose that small weights are less important than large weights, and to use the absolute magnitude of a parameter value as a measure of its importance, under the assumption that the training process naturally forces non-relevant weights into a region around zero.

A major shortcoming of this pruning technique is its weak theoretical motivation. Since parameter estimation is defined in terms of minimization of the error function E, it is natural to seek to use the same error function to find a more principled definition of saliency. We could define the saliency of a model parameter as the change in the error function that results from deletion of that parameter. This could be implemented by direct evaluation, so that for each parameter in the trained network model in turn, the parameter is temporarily set to zero and the error function re-evaluated. Though conceptually attractive, such an approach will be computationally demanding in the case of larger neural spatial interaction models.

We consider instead the change in the error function due to small changes in parameter values (Le Cun, Denker and Solla 1990). If the parameter w_i is changed to $w_i+\delta w_i$, then the corresponding change in the error function E is given by

$$\delta E(\mathbf{w}, S) = \sum_i \frac{\partial E(\mathbf{w}, S)}{\partial w_i}\, \delta w_i + \tfrac{1}{2} \sum_i \sum_j H_{ij} \delta w_i \delta w_j + O(\delta \mathbf{w}^3) \tag{8.27}$$

where H_{ij} denote the elements of the Hessian matrix

$$H_{ij} = \frac{\partial^2 E(\mathbf{w}, S)}{\partial w_i\, \partial w_j} = \nabla^2 E(\mathbf{w}, S) \tag{8.28}$$

If we assume that the parameter estimation process has converged, then the first term in (8.27) will vanish. Le Cun, Denker and Solla (1990) approximate the Hessian by discarding the non-diagonal terms. Neglecting the higher order terms in the expansion then (8.27) reduces to the form

$$\delta E(\mathbf{w}, S) = \tfrac{1}{2} \sum_i H_{ij}\, \delta w_i^2 \,. \tag{8.29}$$

If a parameter with an initial value w_i is set to zero, then the increase in E will be approximately given by (8.29) with $\delta w_i = w_i$, and the saliency values of the weights are given approximately by the quantities $\frac{1}{2} H_{ii}\, w_i^2$ that can be interpreted as a statistical significance measure (see Finoff, Hergert and Zimmermann 1993).

An implementation of this pruning procedure would characteristically consists of the following steps:

- *first*, select a relatively large initial network model [i.e. relatively large *J*],
- *second*, train the network model in the usual way [for example by backpropagation of gradient descent errors] until some stopping criterion is satisfied,
- *third*, compute the second derivatives H_{ij} for each of the parameters, and thus evaluate the saliency weights $\frac{1}{2} H_{ij} \delta w_i^2$
- *fourth*, sort the parameters by saliency and delete some of the low-saliency weights,
- *fifth*, go to the second step and repeat until some overall stopping criterion is satisfied.

Clearly, there are various choices to be made. The most important consideration, however, is to decide upon an appropriate number of parameters to be removed. The choice can be influenced by the number of pruning steps already performed, as well as by visual inspection of the distribution of the test measure (see Finoff, Hergert and Zimmermann 1993). If this problem is solved satisfactorily, the pruning technique which is generally performed interactively, reduces overfitting and improves generalization of neural spatial interaction models.

The *second approach* for directing network architecture modification and selection is through the use of *regularization*, which involves the addition of an extra term $R(\mathbf{w})$ to the error function $E(\mathbf{w})$, designed to penalize mappings that are not smooth. With a sum-of-square error function, the total error function, $E(\mathbf{w}, S)$, to be minimized becomes

$$\widetilde{E}(\mathbf{w}, S) = \sum_{(x_k, y_k) \in S} |\Omega(x_k, \mathbf{w}) - y_k)^2 + \mu R(\mathbf{w}) \tag{8.30}$$

The parameter μ [$\mu \in [0, \infty)$] controls the degree of regularization, i.e. the extent to which the penalty term $R(\mathbf{w})$ influences the form of the solution. Training is performed by minimizing the total error function $\widetilde{E}(\mathbf{w}, S)$ and requires the derivatives of $R(\mathbf{w})$ with respect to the model parameters to be computed efficiently. A spatial interaction function Ω that provides a good fit to the training data will give a small value for $E(\mathbf{w}, S)$, while one that is very smooth will give a small value for $R(\mathbf{w})$. The resulting network model is a compromise between fitting the data and minimizing $R(\mathbf{w})$. One of the simplest regularisers $R(\mathbf{w})$ is called *weight decay*, and consists of the sum of squares of the adaptive model parameters

$$R(\mathbf{w}) = \frac{1}{2} \sum_i w_i^2 \tag{8.31}$$

the first derivative of which leads to weight decay in the weight updates (see Hanson and Pratt 1989). The use of this form of regulariser corresponds to ridge regression in conventional curve fitting. Hinton (1987) has shown empirically that such a regulariser can lead to significant improvements in network generalization.

One of the difficulties of the simple regulariser (8.31) is that it tends to favour many small parameter values rather than a few large ones. This problem can be overcome by using a modified penalty term of the form

$$R(w) = \sum_i \frac{w_i^2}{\hat{w}^2 + w_i^2}. \tag{8.32}$$

where $\hat{w}$ corresponds to a parameter that sets a scale usually chosen by hand to be of order unity. Use of this penalty term has been called *weight elimination* (Weigend, Huberman and Rumelhart 1990). It will tend to favour a few large parameter values rather than many small ones, and is thus more likely to eliminate parameters from the model than (8.31). This leads to a form of network model pruning which is combined with the training process itself, rather than alternating with it, as in the case of pruning techniques.

The principal alternative to regularization and weight pruning as a way of controlling the effective complexity of a neural spatial interaction model is the procedure of *stopped* or *cross-validation* training. This method, in which an oversized network model is trained until the error on a further validation data set deteriorates, is a true innovation deriving from neural network research, since model selection does not require convergence of the training process. The training process is used here to perform a directed search of the parameter space for a model that does not overfit the data, and thus has superior generalization performance. Various theoretical and empirical results have provided strong evidence for efficiency of cross-validation training (Weigend, Rumelhart and Huberman 1991a; Baldi and Chauvin 1991; Finnoff 1991; Fischer and Gopal 1994a, b). Although many questions remain, a picture is starting to emerge as to the mechanisms responsible for the effectiveness of this procedure. In particular, it has been shown that stopped training has the same sort of regularization effect [i.e. reducing model variance at the cost of bias] that penalty terms provide.

Since the efficiency of a model selection technique may depend on the data fitting problem at hand, a truly informative comparison of performance can only be obtained by testing on a wide range of examples containing varying degrees of complexity.

8.6 Evaluating the Generalization Performance of a Neural Spatial Interaction Model

The third stage in the model building process corresponds to testing the out-of-sample [generalization] performance of the model. First, we consider the standard approach, and then evaluation by means of bootstrapping.

The Standard Approach

The standard approach for finding a good neural spatial interaction model [see Fischer and Gopal 1994a] is to split the available set of samples into three sets: training, validation, and test sets. The training set is used for parameter estimation. In order to avoid overfitting, a common procedure is to use a network model with sufficiently large J for the task, to monitor – during training – the out-of-sample performance on a separate validation set, and finally to choose the network model that corresponds to the minimum on the validation set, and employ it for future purposes such as the evaluation on the test set. It has been frequent practice in the neural network community to fix these sets. But recent experience has found this approach very sensitive to the specific splitting of the data. Thus, the usual tests of out-of-sample or generalization reliability may appear over optimistic.

Randomness enters into neural spatial interaction modelling in two ways: firstly in the splitting of data samples, and secondly in choices about the parameter initialization and control parameters of the estimation approach used [such as κ in the global search procedure described in Section 8.4]. This leaves one question wide open. What is the variation in out-of-sample performance as one varies training, validation and test sets? This is an important question, since real world problems do not come with a tag on each pattern telling how it should be used. Thus, it is useful to vary both the data partitions and parameter initializations to find out more about the distributions of out-of-sample errors.

Monte Carlo experiments can provide certain limited information on the behaviour of the test statistics. The limitation of Monte Carlo experiments is that any results obtained pertain only to the environment in which the experiments are carried out. In particular, the data-generating mechanism has to be specified, and it is often difficult to know whether any given data-generating mechanism is to any degree representative for the empirical setting under study.

Evaluation by Means of Bootstrapping

Motivated by the desire to obtain distributional results for the test statistics, that rely neither on large size approximations nor on artificial data generating assumptions, statisticians have developed resampling techniques such as bootstrapping that permit a fairly accurate estimation of finite sample distributions for test statistics of interest. Bootstrapping is a computer intensive non-parametric approach to statistical inference that enables the estimation of standard errors by re-sampling the data in a suitable way (see Efron and Tibshirani 1993). This idea can be applied to the third stage in neural spatial interaction modelling in two different ways. One can consider each input-output pattern as a sampling unit, and sample with replacement from the input-output pairs to create a bootstrap sample. This is sometimes called boostrapping pairs (Efron and Tibshirani 1993) since the input-output pairs remain intact and are resampled as full patterns. On the other hand, one can consider the predictors as fixed, treat the model residuals $y_k - \Omega(x_k, \hat{w})$ as the sampling units, and create a bootstrap sample by adding residuals to the model fit $\Omega(x_k, \hat{w})$. This is termed the *bootstrap residual approach.* In this approach, the residuals obtained from one specific model are used in rebuilding patterns to obtain error bars reflecting all sources of error, including model misspecification. If interest is in discovering variation due to data samples rather than error bars, then the *bootstrapping pairs approach* is more appropriate.

The bootstrapping approach is summarized in Fig. 8.2. The idea behind this approach is to generate many pseudo-replicates of the training, validation and test sets, then re-estimate the model parameters $\boldsymbol{w}$ on each training bootstrap sample, and test the out-of-sample performance on the test bootstrap samples. In this bootstrap world, the errors of forecast, and the errors in the parameter estimates, are directly observable. The Monte-Carlo distribution of such errors can be used to approximate the distribution of the unobservable errors in the real parameter estimates and the real forecasts. This approximation is the bootstrap – it gives a measure of the statistical uncertainty in the parameter estimates and the forecasts. The focus in the third stage of model building is on the performance of the forecasts measured in terms of *ARV*.

In more detail, the approach may be described by the following steps: The details may be rather complicated, but the main idea is straightforward:

Step 1: Conduct three totally independent resampling operations in which

(i) B independent *training bootstrap samples* are generated, by randomly sampling $K1$ times, with replacement, from the observed input-output pairs $S = \{(x_1, y_1), (x_2, y_2), ..., (x_K, y_K)\}$:

$$S_{Train}^{*b} = \{(x_{k_1}^{*b}, y_{k_1}^{*b}), (x_{k_2}^{*b}, y_{k_2}^{*b}), ..., (x_{k_{K1}}^{*b}, y_{k_{K1}}^{*b})\} \quad (8.33)$$

for $k_1, k_2, ..., k_{K1}$ a random sample of integers 1 through $K1 < K$ and $b = 1, ..., B$

(ii) B independent *validation bootstrap samples* are formed, by randomly sampling $K2$ times, with replacement, from the observed input-output pairs S:

$$S_{Valid}^{*b} = \{(x_{k_1}^{*b}, y_{k_1}^{*b}), (x_{k_2}^{*b}, y_{k_2}^{*b}), ..., (x_{k_{K2}}^{*b}, y_{k_{K2}}^{*b})\} \quad (8.34)$$

for $k_1, k_2, ..., k_{K2}$ a random sample of integers 1 through $K2 < K1 < K$ and $b = 1, ..., B$

(iii) B independent *test bootstrap samples* are formed, by randomly sampling $K3$ times, with replacement, from the observed input-output pairs S:

$$S_{Test}^{*b} = \{(x_{k_1}^{*b}, y_{k_1}^{*b}), (x_{k_2}^{*b}, y_{k_2}^{*b}), ..., (x_{k_{K3}}^{*b}, y_{k_{K3}}^{*b})\} \tag{8.35}$$

for $k_1, k_2, ..., k_{K3}$ a random sample of integers 1 through $K3 = K2 < K1 < K$ and $b = 1, ..., B$

Step 2: For each training bootstrap sample S_{Train}^{*b} (b=1, ..., B) minimize

$$\sum_{k=1}^{K1} \left| y_k^{*b} - \Omega_L(x_k^{*b}, w) \right|^2 \tag{8.36}$$

with an appropriate parameter estimation procedure. During the training process the ARV performance of the model is monitored on the bootstrap validation set. The training process is stopped when ARV^* (S^{*b}_{valid}) starts to increase. This yields bootstrap parameter estimates $\hat{w}^{*b}$.

Step 3: Calculate the bootstrap ARV-Statistic of generalization performance, $\widehat{ARV}^*$ (S^{*b}_{Test}), for each test bootstrap sample. The distribution of the pseudo-errors $ARV^* - \widehat{ARV}$ can be computed, and used to approximate the distribution of the real errors $\widehat{ARV}^* - ARV$. This approximation is the bootstrap.

Step 4: The variability of $\widehat{ARV}^*$ (S^{*b}_{Test}), for b=1, ..., B gives an estimate of the expected accuracy of the model performance. Thus, estimate the standard error of the generalization performance statistic by the sample standard deviation of the B bootstrap replications:

$$\hat{\text{se}}_B = \left\{ \sum_{b=1}^{B} [\widehat{ARV}^*(S_{Test}^{*b}) - \widehat{ARV}^*(.)]^2 / (B-1) \right\}^{\frac{1}{2}} \tag{8.37}$$

where

$$\widehat{ARV}^*(.) = \sum_{b=1}^{B} \widehat{ARV}^*(S_{Test}^{*b}) \, . \tag{8.38}$$

Note that the name 'bootstrap' refers to the use of the original sample pairs to generate new data sets. The procedure requires retraining of the neural spatial interaction model B times [see *Step 2*]. Typically, B is in the range $20 \le B \le 200$ (see Tibshirani 1996). Increasing B further does not bring a substantial reduction in variance (Efron and Tibshirani 1993).

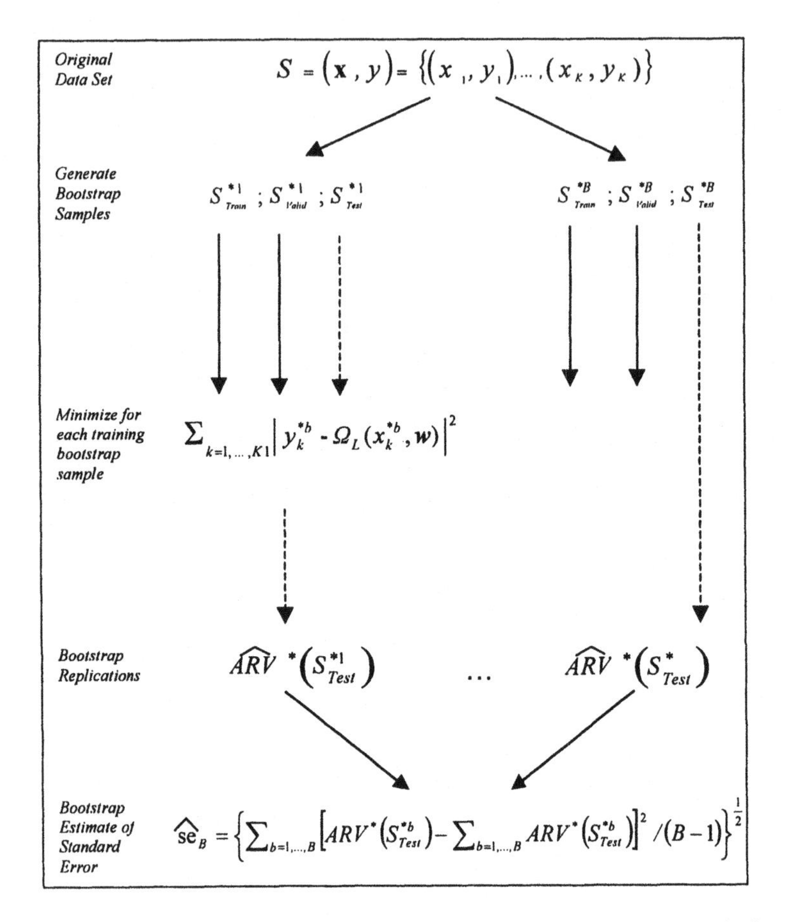

Fig. 8.2 A diagram of the bootstrap procedure for estimating the standard error of the generalization performance of neural spatial interaction models

8.7 Conclusion and Outlook

The explosion of interest in neural network research has been accompanied by no little measure of hyperbole concerning the potential of computational neural networks. In addition, there is a certain mystique perceived by those outside the neural network community arising from the origins of computational neural networks in the study of natural neural systems, and in the associated metaphorical jargon of the field. Unfortunately, both the hyperbole and the mystique have had the effect of limiting the amount of serious attention given to computational neural network modelling in spatial analysis.

Neural spatial interaction models are attractive non-linear models. But despite all the progress made in the past few there are several areas that need further research. For example, finding a suitable neural spatial interaction model for a particular application is still a challenging task. It is possible to experiment with handcrafted designs found by trial and error procedures. But a much more appealing approach would be to employ techniques that tune the network topology automatically. Suggestions for work along these lines have already begun appearing. Fischer and Leung (1998), for example, suggest combining neural network topology optimization performed by genetic algorithms with gradient descent backpropagation for parameter estimation. The use of evolutionary programming rather than genetic algorithms for model selection is an area open to investigation.

Up to now research activities had focused exclusively on unconstrained neural spatial interaction modelling. In many practical situations we have, in addition to (x_k, y_k)-data samples, information about the inflow and/or outflow totals which the mapping should satisfy. This is referred to as prior knowledge, and its inclusion in the network model design process can often lead to substantial improvements in performance. In principle, prior knowledge can be incorporated into the network structure itself or into the pre-processing and post-processing stages. It can also be used to modify the training process through the use of regularization. Further research is nevertheless necessary to identify a generic way of handling the current constraints in neural spatial interaction modelling.

Acknowledgement: The author gratefully acknowledges grant no. P12681-INF provided by the Fonds zur Förderung der wissenschaftlichen Forschung (FWF).

9 A Neural Network Approach for Mobility Panel Analysis

Günter Haag
Steinbeis Transfer Centre, Applied Systems Analysis

9.1 Introduction

Currently, neural networks are being widely discussed in many research areas. But despite this clear interest, it is curious to find such a large gap between theoretically based research activities and real world applications. Especially in the field of social sciences, worthwhile applications of computational neural networks [CNNs] are the exception rather than the rule.

It was the aim of the present research project to apply and compare different methods and models from the field of econometric modelling and neural networks to the analysis of the German Mobility Panel [GMP]. The Mobility Panel was set up in 1994 by the German Ministry of Transport in order to collect data on individual mobility behaviour. The resulting information has been used as the basis for this pilot study, financed by the German Bundesministerium für Verkehr, Bau- und Wohnungswesen (Hautzinger et al. 2000).

Two different types of variable have been analyzed. First, a continuous variable called DAU_SUM, namely the overall time used on average for mobility purposes by an individual per day [in minutes]. Second, a discrete [binary] variable called NUTZPKW indicating the use [or not] of a car by a specific person on a particular day of the week. The analysis of DAU_SUM was performed via a multiplicative econometric panel model, as well as a feedforward neural network with a modified backpropagation learning rule in combination with different pruning procedures. The variable NUTZPKW was analyzed via a probit-model and again a feedforward neural network with a modified backpropagation learning rule in combination with different pruning procedures.

A neural network model with feedback loops can be seen as a dynamic system. From this point of view, a given input signal may disturb the dynamics of the system. The network then performs cycles until it settles down into another stable state (Baldi and Hornik 1989). If the network approaches a global minimum state, it will essentially generate the lowest energy or cost function for the given problem. The weights are then information containing points. This kind of mechanism can therefore be used for associative memory problems (Weigend, Huberman and Rumelhart 1990).

The processing capabilities differ widely among the various types of neural networks. Some CNNs are appropriate for pattern matching and classification, while others can handle constraint satisfaction or optimization problems (for example, see Hruschka and Natter 1992; Poggio and Girosi 1989). One of the challenges faced by model builders is to determine which type of network is appropriate for a given problem. It is here that experience in model building, design and application of neural networks is of crucial importance (Fischer 1998b; Hugo 1998).

In this study, we needed to find a neural network approach for the analysis and simulation of the GMB which would capture several different properties. First of all, it had to be able to take into account the specific character of the panel data, i.e. that the mobility behaviour of one and the same person was considered for different days of the week and for different years. In other words, the neural net should be able to identify an individual person, given certain characteristics of that person. Secondly, in the case of the variable NUTZPKW, the selected neural network had to be able to classify an individual as a car user or not a car user. Thirdly, for the variable DAU_SUM, the corresponding neural network had to perform a function approximation task depending on the characteristics of the various individuals.

In relation to the use of neural networks, some words of caution are necessary. The steadily increasing attention over the last decade to neural networks has created considerable enthusiasm. But though they have proven themselves capable of solving hard problems, they cannot perform magic! As with any model system, thorough functional specifications and test plans are needed to ensure a reliable application. Of course, considering the analysis of panel data with neural networks also requires a multidisciplinary interaction of specialized scientists.

9.2 The German Mobility Panel

The German Mobility Panel contains relevant socio-demographic and travel data [following the German KONTIV-design] of a representative sample of people in West Germany aged 10 years and above. Data were collected during the first wave [in autumn] over a period of seven days. Most people participated in three waves to avoid panel conditioning. The data collection was undertaken with the help of a travel diary and questionnaires (Zumkeller 1994; Chlond, Lipps and Zumkeller 1996).

At the household level, small households without cars are underrepresented. On the individual level, this meant that older women were underrepresented. Household and person distortions are weighted by car ownership, household size, sex and age, respectively. Until now, the sample sizes [in terms of cohorts] shown in Table 9.1 have been involved.

A selective, but insignificant, drop out can be observed for younger persons. Data corrections have been performed with the aid of the software tool GRADIV [developed by the Institute for Transport Studies of the University of Karlsruhe], which allows visualization of the data reported on an individual level. An example of a trip home on Sunday afternoon which went unreported is shown in the following Fig. 9.1 [thick bars: trips; thin bars: activities].

Table 9.1 The sample sizes in terms of different cohorts. Sample development, participants [sum], new participants [bold] and repeaters [persons and percentage of persons with respect to the previous year]

Cohort [Year]	1994 Number	1995 Number [%]	1996 Number [%]	1997 Number [%]	1998 Number [%]	1999 Number [%]	Sum
1994	**517**	295 (57)	149 (51)				961
1995		**449**	197 (44)	146 (74)			792
1996			**1,141**	846 (74)	633 (75)		2,620
1997				**529**	346 (69)	296 (86)	1,171
1998					**503**	394 (78)	897
1999						**1,242**	1,242
Sum	517	744	1,487	1,521	1,482	1,887	7,683

In terms of attrition or fatigue within a wave, apart from minor effects during the pilot waves in 1994/1995, no significant effects can be detected. This shows the high data quality of the panel. As a result of attrition or mortality between two waves, a slightly smaller number of trips are reported in the second wave. This reduction does not hold for distance and time budgets. By a suitable weighting of the data, this effect can be reduced. Moreover, for the different waves, slightly varying trip distance distributions can be observed. This effect is due to minor variations in the sample designs affecting short distance trips and because of the relatively small total sample size [this affects long distance trips, which were a rare event]. These effects can be cancelled by means of a combined trip distance distribution calculated as a weighted sum of the last four trip length distributions.

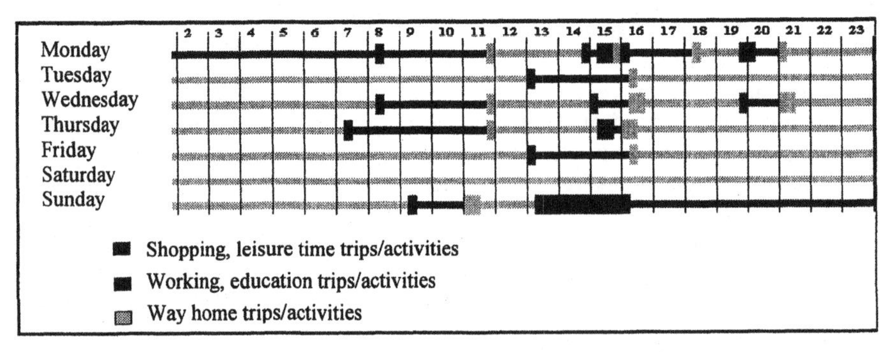

Fig. 9.1 Example of a trip home on Sunday afternoon not reported

In order to compensate for the lack of seasonal effects and, consequently, the underrepresentation long distance trips, a new panel with only long distance trips made over a longer time span was established in 2000. The data collection of this survey too was carried out by Infratest [Munich], and the scientific supervision conducted by the Institute for Transport Studies of the University of Karlsruhe.

9.3 Classical Panel Analysis

The German Mobility Panel was analyzed using different model paradigms. One of the main aims of the research was to compare the application of econometric panel data models with the use of neural networks. This section provides some background to the econometric fixed effects panel model used (Baltagi 1995; Hsiao 1986). In contrast to the random effects model, individual variations here are captured in individual constants and not in an individual specific error term. This provides easy calculation. The model is defined as:

$$Y_{it} = \alpha_i + \beta x_{it} + \varepsilon_{it} \tag{9.1}$$

for the i-th individual ($i = 1, ..., n$, $t = 1, ..., T$). α_i represent person-specific time-invariant parameters, ε_{it} is a general error term, and β represents the influence of the input matrix x_i on the model output Y_{it}. This model (9.1) was applied to the analysis of the continuous variable LOG[DAU_SUM]. For the analysis of the binary variable NUTZPKW a probit-model with random effects was used.

9.4 Application of Computational Neural Networks to the German Mobility Panel

Instead of following the classical panel analysis approach, it was decided to investigate whether or not a neural network would be able to recognize [classify] an individual person if a subset of characteristics of that particular person was known, e.g. sex, age, place of work, etc. It should be possible for the network to capture and estimate individual variations from the data set without identifying each individual separately. If so, the number of input units could be reduced dramatically. In this case, the structure of the CNN could be very simple in its architecture (see Fig. 9.2) and the method provide an interesting alternative to classical panel methods.

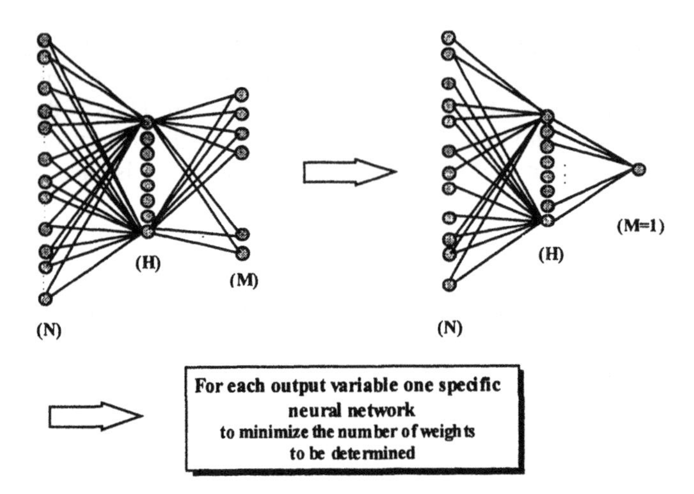

Fig. 9.2 The general structure of feedforward neural networks

An important experience gained in this research study concerns the structure of the neural network. It appeared to be justified, and supported by the principle of minimizing the number of weights in relation to the data available, to use one CNN for each output task to be analyzed. In other words, we could use two networks, one for the variable LOG[DAU_SUM] and one for NUTZPKW (see Fig. 9.3). The great advantage of this procedure – using different networks for different variables, even if the same input vector was used – consists in the considerable reduction of the number of weights to be determined per network. This is based on the fact that after the pruning procedure for each variable, DAU_SUM and NUTZPKW, different input variables may be relevant and therefore selected.

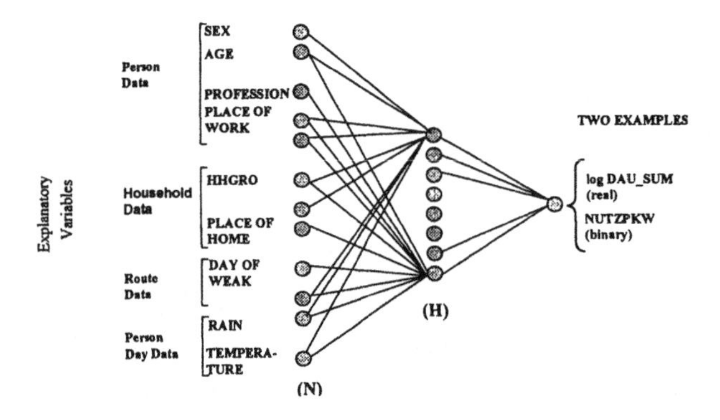

Fig. 9.3 The structure of neural networks for the analysis of the GMP

Fig. 9.3 illustrates the structure of the feedforward network with backpropagation learning algorithm used for the analysis of the Mobility Panel data. The transfer functions of the hidden and output units are assumed to be logistic functions and identical with respect to the parameter values. The number H of hidden units is set to $H < 0.3\ N$, with N being the number of input units, to obtain a sufficient compression of information. In the application, $H = 21$ (20) has been tested to be appropriate. The architecture for the analysis of the variables DAU_SUM or LOG[DAU_SUM] and NUTZPKW therefore reads (N-H-1). Therefore, without pruning, this means that at the beginning of the pruning process we have $N\ (H+1)$ connections [weights] to be determined by the learning process.

For the task of parameter estimation, we utilize the error backpropagation technique, a recursive learning procedure using the gradient descent procedure to minimize the least-square error function. We used the off-line training approach. Input signals were presented in random order and the parameter updates [weight adjustments] following the momentum update rule, according to Rumelhart and McClelland (1986) and Rumelhart, Hinton and Williams (1986a, 1986b)

$$\Delta w_{ij}^{(l)}(t+1) = -\eta \frac{\partial E(w)}{\partial w_{ij}^{(l)}} + \alpha\ \Delta w_{ij}^{(l)}(t) \tag{9.2}$$

where $\partial E(w)/\partial w_{ij}^{(l)}$ denotes the partial derivative of E with respect to the parameter $w_{ij}^{(l)}$ [the error gradient], η represents the learning rate and α the momentum parameter ($0 \leq \alpha \leq 1$). The index (l) relates to the different layers of the network and t indicates the subsequent time steps of the weight adaptation or learning process. For more details of the parameter estimation process see Chapter 2. The network parameters used are summarized in Table 9.2.

Table 9.2 The general network parameters used in the training process

General Parameters of the Network and Learning Rule		
Learning Rate	η	0.01
Momentum Parameter	a	0.90
Activation Function	$g(h) = 1/(1 + \exp(-\beta h))$	Logistic function
Pseudo Temperature	ß	1.00

In order to determine the 'relevant' or explanatory [input] variables for a given output variable, unimportant connections [weights] have to be removed using an appropriate pruning procedure (see also Chapter 2). It is possible to prune unimportant weights [link pruning] or even units [unit pruning] (Sietsma and Dow 1988). After pruning, the network has to be retrained in order to adjust the weights to the new architecture. Different types of pruning procedures, such as magnitude based pruning [MBP], optimal brain damage [OBD] (Le Cun, Denker and Solla 1990), optimal brain surgeon [OBS] or skeletonization [SKEL] have been widely applied and discussed (Zimmermann 1994; Reed 1993; Mozer and Smolensky 1989).

A systematic simplification of the neural networks for the analysis of the GMP data can be obtained by unit pruning applied to both the input array and the hidden layer. Skeletonization has the advantage of leading to a clear and easily interpreted network architecture, since all connections [weights] starting or ending at a particular unit will be cancelled simultaneously. Skeletonization has been tested and selected as an appropriate pruning procedure, especially for socio-economic data.

The pruning procedure starts after a first training phase of 1,000 epochs. Further pruning steps occur after every 200 iterations, as long as the network errors are decreasing. An irreversible increase of the network error *ARV* [average relative variance] stops the pruning procedure. *ARV* is defined as

$$ARV_{adj} = 1 - R^2_{adj} = \frac{Var(\varepsilon)}{Var(y)} \cdot \frac{(N-1)}{(N-k-1)} \tag{9.3}$$

where ε is the residual error and y the observed value of the predicted variable, N is the number of observations and k the degrees of freedom [number of weights].

The Preprocessing of Data

One of the aims of this section is to demonstrate how a neural network is specified with respect to the data base, and how its final architecture [number of units, number of connections] are established for a real world application. The former task leads to the preprocessing problem, namely an adequate scaling of the data [input and output data] as required by the transfer functions. Choosing how to transform the data to be presented to the neural network is as important as choosing what kind of data to use. Poorly represented data can be very difficult, if not impossible, for a network to train on. A design feature of most neural network algorithms is that they require the data to be scaled to a small range, such as 0.0 to 1.0 (Zell 1994; von Stackelberg 1998; Müller and Haag 1996).

Table 9.3 Scaling of the number of cars per household

Number of Cars per Household	Input Value [real]
0	0.00
1	0.25
2	0.50
3	0.75
4 or more	1.00

Questions concerning the database which can be answered by YES or NO or data like sex [man or woman] can be represented simply by binary values (0/1) and are therefore easily transferable to this range and are represented by a one-bit binary unit only. Values such as the number of cars per household, the size of the household etc. are scaled according to Table 9.3 and represented by one real-valued unit.

A 1-of-N-Code is used to represent most of the variables of the German Mobility Panel, such as profession, place of work, place of home, age, day of the week, type of household, and so on. In practice, the problem of missing data is of crucial importance. If some data are missing for a specific person, it must be possible for the network to recognize it. The coding of missing data can be performed easily in case of 1-of-N-Code as shown in Table 9.4.

Table 9.4 Scaling of the shopping facility variable [1-of-N-Code]

Shopping Facility in the Neighbourhood	1-of-2-Code [2 binary units]
YES	1 0
NO	0 1
No Comment	0 0

In Fig. 9.4 the different coding of panel data is demonstrated for two input variables: place of work [LAGEARB] and job qualification [SCHULAB]. With respect to the different attributes of these variables, the place of work can be coded by one real value, and for job qualification a 1-of-5-Code seems to be adequate.

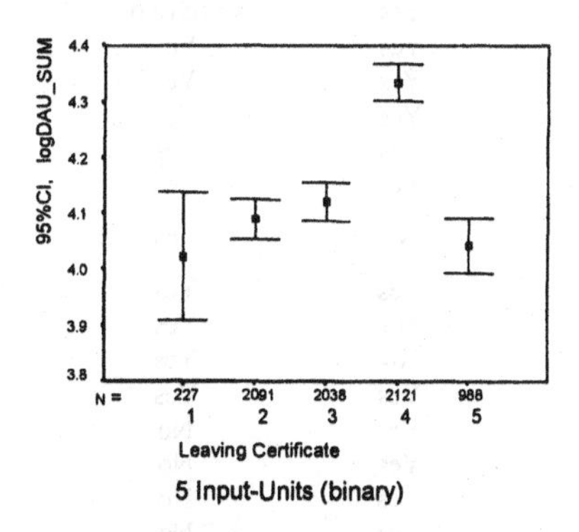

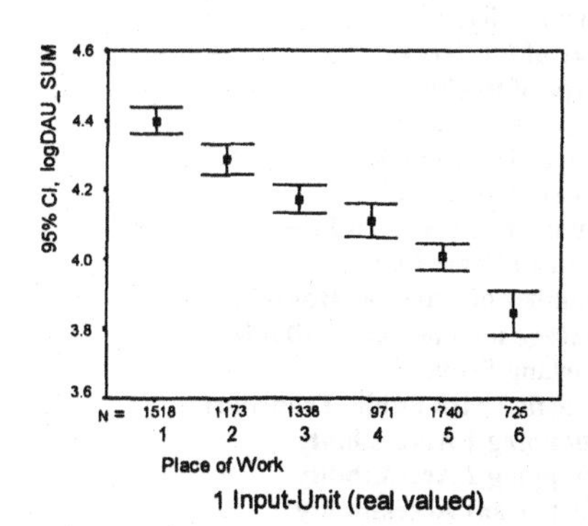

Note: For different attributes the range of variation of the output variable LOG[DAU_SUM] has been depicted

Fig. 9.4 The coding problem: Two different possibilities

9.5 Analysis of the Variable LOG[DAU_SUM]

For the analysis of the variable DAU_SUM, i.e. the average time per person and per day spent on mobility, the Mobility Panel data for 1994 to 1997 was used. Only data belonging to a 'normal' day, without special activities and where at least one trip was performed, were selected from this data set. In order to include panel specific details, it was further required that only persons who had participated in at least two waves were selected. This led to 13,099 observations being used in the analysis. Furthermore, instead of the independent variable DAU_SUM, its logarithm LOG[DAU_SUM] was used. This was to make it comparable with the classical panel analysis, which focused on the variable LOG [DAU_SUM] only. From the neural network point of view, analysis of DAU_SUM would have been preferable, since a much better intuitive interpretation of this variable can be given.

Table 9.5 The different input variables of the two CNNs for LOG[DAU_SUM] and NUTZPKW

Input Variable	Number of Units	Log[DAU_SUM]	NUTZPKW
Year (1994 – 1997)	4	Yes	Yes
Sex	1	Yes	Yes
Leaving Certificate	5	Yes	Yes
Place of Work	6	Yes	Yes
Profession	7	Yes	Yes
Public Transport	5	Yes	Yes
Parking Problems	4	No	Yes
Car Availability	3	No	Yes
Season Ticket	2	No	Yes
Bicycle, Motor Cycle	2	Yes	No
Age	7	Yes	Yes
Driver License	1	Yes	All have one
Day of the Week	7	Yes	Yes
Type of Region	1	Yes	Yes
Type of Household	4	Yes	Yes
Place of Household	6	Yes	Yes
Size of Household	1	Yes	Yes
Number of Children Less than 10 Years Old	1	No	Yes
Number of Cars per Household	1	Yes	Yes
Garage for the Car Available	1	No	Yes
Parking Problems	4	No	Yes
Satisfied with Public Transport	2	Yes	Yes
Shopping 1 Accessibility	2	Yes	No
Shopping 2 Accessibility	2	Yes	No
Restaurant Accessibility	2	Yes	No
Cinema Accessibility	2	Yes	No
Sport Facilities	2	Yes	No
Mobility Type	4	Yes	No
Type of Trip	3	No	Yes
Rain	2	Yes	Yes
Temperature	2	No	Yes
		76 Input Units	75 Input Units

Different neural networks, with respect to the number of input units, the kind of data preprocessing and the number of hidden units were tested and evaluated. Finally, a feedforward network (76-21-1) was selected. Table 9.5 shows a list of the input variables used in this network for LOG[DAU_SUM] and NUTZPKW. The number of units used for the coding of the input and output variables are also indicated. This network provided the starting point in a skeletonization pruning procedure (see Fig. 9.5).

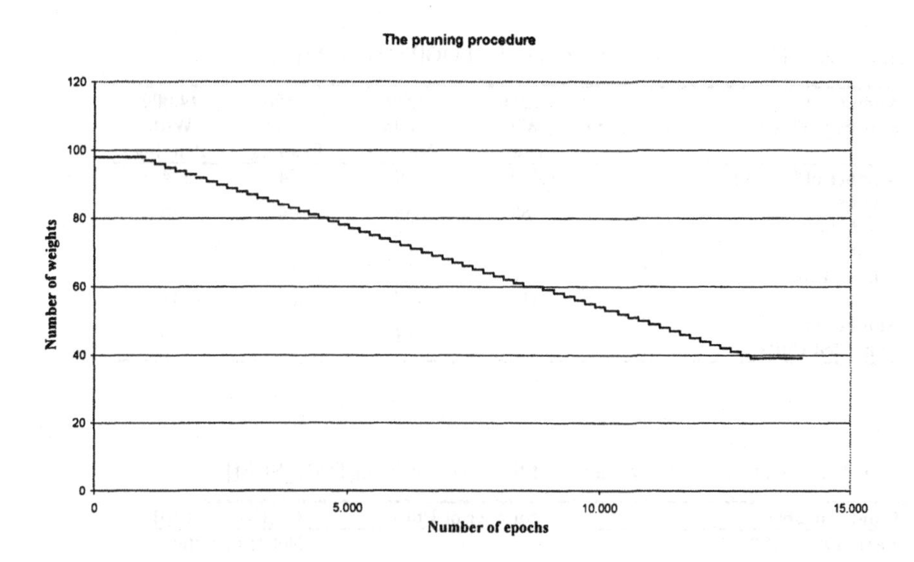

Fig. 9.5 The pruning procedure, showing the number of weights [links] in relation to the number of learning steps [epochs] with skeletonization pruning

Table 9.6 shows the results of the pruning procedure. It can be seen, as expected, that with an increasing number of pruning steps, the performance quality of the CNN decreases. Comparing the net performance $R^2 = 0.460$ of the neural network with the results obtained via a classical panel analysis $R^2 = 0.388$, it is evident that the trained network provides a high value of net performance with respect to the socio-economic data. This indicates that a neural network is able to identify individual persons, even if only subsets of an individual's characteristics are presented to the network.

The cancelled input units after certain pruning steps indicate the variables or certain attributes of variables which seem to be irrelevant as explanatory variables of LOG[DAU_SUM]. Therefore, the variables remaining after a number of pruning steps [7,500 epochs] are considered to be explanatory [relevant] variables

In Table 9.7 the importance of the different input variables of the CNN for LOG[DAU_SUM] are shown. The number and type of input units define the input array of the CNN. Even in case of unit pruning, the pruning procedure does not usually delete all units related to specific input variable simultaneously, but only some of the attributes belonging to certain variables. Those variables which are only partially deleted are indicated in Table 9.7 as *not so important*. This information can also be used to simplify the questionnaire of the panel. Here one further advantage of neural nets becomes obvious. A CNN adaptively estimates mathematical functions from data without specifying mathematically how outputs depend on inputs. In other words, a neural network provides an adaptive model-free selection of relevant input variables.

Table 9.6 Pruning and net performance for LOG[DAU_SUM]

Number of Epochs [trained]	7,500 No Pruning	2,500 With Pruning	4,500 With Pruning	7,500 With Pruning	14,000 With Pruning
Number of Weights	1,617	1,449	1,239	924	324
Number of Input Units	76	68	58	44	26
Number of Hidden Units	21	21	21	21	12
R^2	0.460	0.411	0.407	0.366	0.143
Number of Cancelled Units	No	8	18	32	50

Table 9.7 Relevant input variables of the CNN for LOG[DAU_SUM]

Input Variable	Number of Units	Log[DAU_SUM]
Year (1994 – 1997)	4	Not so important
Sex	1	Very important
Leaving Certificate	5	Important
Place of Work	6	Not so important
Profession	7	Full time job, important
Public Transport	5	Not so important
Bicycle, Motor Cycle	2	Not so important
Age	7	Important
Driver License	1	Very important
Day of the Week	7	Important
Type of Region	1	Very important
Type of Household	4	Important
Place of Household	6	Not so important
Size of Household	1	Very important
Number of Cars per Household	1	Important
Satisfied with Public Transport	2	Very important
Shopping 1 Accessibility	2	Important
Shopping 2 Accessibility	2	Very important
Restaurant Accessibility	2	Not important
Cinema Accessibility	2	Important
Sport Facilities	2	Very important
Mobility Type	4	Not important
Rain	2	Important
		76 Input Units

Since it is the aim of this contribution to demonstrate the application of neural networks to a huge data set [the German Mobility Panel], we shall skip further discussion. In Hautzinger et al. (2000) a detailed interpretation of the results of the analysis is given, as well as some simulations using the trained neural network. For policy considerations, it is crucial to determine the dependence of the variables, e.g. DAU_SUM, on different scenarios. A scenario in this context means a correspondingly modified input pattern [scenario input pattern].

The Mobility Panel contains detailed information about the spatial location of households, place of work, and the transportation network. In order to determine the impact of the spatial location of a household on the overall time spent for mobility purposes, the individual data records are presented to the trained net where a specific spatial location is defaulted and the corresponding value of DAU_SUM determined via simulation.

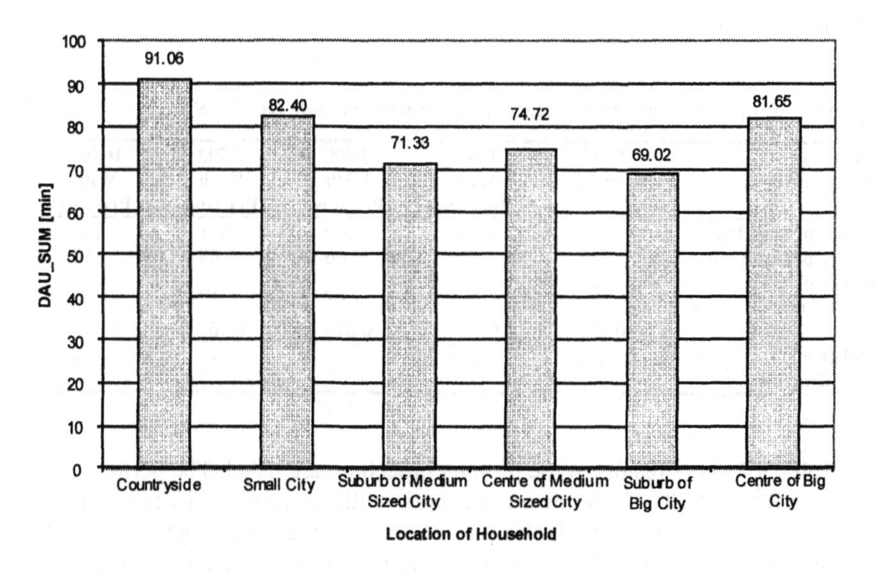

Fig. 9.6 Dependence of the overall time spent for mobility purposes on the spatial location of the households

Fig. 9.6 shows the dependence of the overall time spent for mobility purposes on the spatial location of the households. It can be seen that the spatial location of a household has a clear effect on the output variable DAU_SUM. As might be expected, the average daily time spent for mobility purposes is much higher for households living in the centre of a big city [81 min.] and in the countryside [91 min.], than those in the suburbs of big cities [69 min.] and medium-sized cities [71 min].

9.6 Analysis of the Variable NUTZPKW

In the next section, a discrete [binary] variable called NUTZPKW will be analysed. This variable indicates whether or not a car was used by a specific person on a particular day of the week. If this is the case, the variable NUTZPKW is set to 1, otherwise NUTZPKW is set to 0. For this variable, the data base consisted of 9,067 observations. Altogether 6,207 individuals used a car and 2,860 did not use a car on a particular day of the week.

The set of input variables used is listed in Table 9.5. To test against overfitting and to select an optimal network, different neural network specifications were tested and evaluated. The resulting feedforward network [75-20-1] provided the starting point in a skeletonization pruning procedure. If the CNN provided an output value > 0.5 for certain input data, the variable NUTZPKW was set to 1, otherwise NUTZPKW = 0.

Table 9.8 Pruning and net performance for the binary variable NUTZPKW

Number of Epochs [trained]	**5,000 No Pruning**	**1,000 With Pruning**	**3,000 With Pruning**	**7,500 With Pruning**	**10,000 With Pruning**
Number of Weights	1,520	1,520	1,340	931	644
Number of Input Units	75	75	66	48	45
Number of Hidden Units	20	20	20	19	15
R^2	0.791	0.767	0.674	0.467	0.337
Number of Cancelled Units	No	No	9	27	30

In Table 9.8 some results of the pruning procedure are listed. The performance parameter R^2 = 0.79 of the neural network without pruning is rather high. However, in case of binary variables the *sensitivity* – the percentage of correctly identified car users – and the *specitivity* – the percentage of correctly identified non car users – are more appropriate performance indicators.

Table 9.9 shows the importance of the different input variables of the CNN for NUTZPKW. However, since the pruning procedure does not delete as many links as in Table 9.7, the question whether ot not to use the car on a particular day seems to be based on a rather complicated interlinked decision process.

In Table 9.10 the cross table based on the CNN for NUTZPKW is given. From all 6,207 car users, only 140 [1.5 percent] were falsely classified. The classification error for non car users on a particular day is 3.4 percent [312 of 2,860 non car users]. This is quite an impressive result. There were in all 95.0 percent correctly identified observations.

Table 9.9 The relevant input variables of the CNN for NUTZPKW

Input Variable	Number of Units	NUTZPKW
Year (1994 – 1997)	4	Important
Sex	1	Very important
Leaving Certificate	5	Important
Place of Work	6	Important
Profession	7	Important
Public Transport	5	Not so important
Parking Problems	4	Not so important
Car Availability	3	Important
Season Ticket	2	Very important
Age	7	Important
Day of the Week	7	Not so important
Type of Region	1	Important
Type of Household	4	Important
Place of Household	6	Not so important
Size of Household	1	Important
Number of Children Less than 10 Years Old	1	Very important
Number of Cars per Household	1	Very important
Garage for the Car	1	Not relevant
Parking Problems	4	Not relevant
Satisfied with Public Transport	2	Very important
Type of Trip	3	Very important
Rain	2	Not so important
Temperature	2	Not so important
		75 Input Units

Table 9.10 Cross table for the results of the CNN for the output variable NUTZPKW

Observed Values	Value of the Model 0	Value of the Model 1	Sum
0	2,548	312	2,860
1	140	6,067	6,207
Sum	2,688	6,379	9,067

The distribution of the real output values of the CNN for NUTZPKW is presented in Fig. 9.7. The strong concentration of this bimodal distribution shows that, for given individual input data, the classification *car user/non car user* by the criteria that the estimated output value > 0.5 (< 0.5) is well justified. This result can be compared with a panel model developed by Hautzinger et al. (2000) shown in Table 9.11.

Table 9.11 Cross table for the results of a probit model for the output variable NUTZPKW

Observed Values	Value of the Model 0	Value of the Model 1	Sum
0	1,444	1,416	2,860
1	678	5,529	6,207
Sum	2,122	6,945	9,067

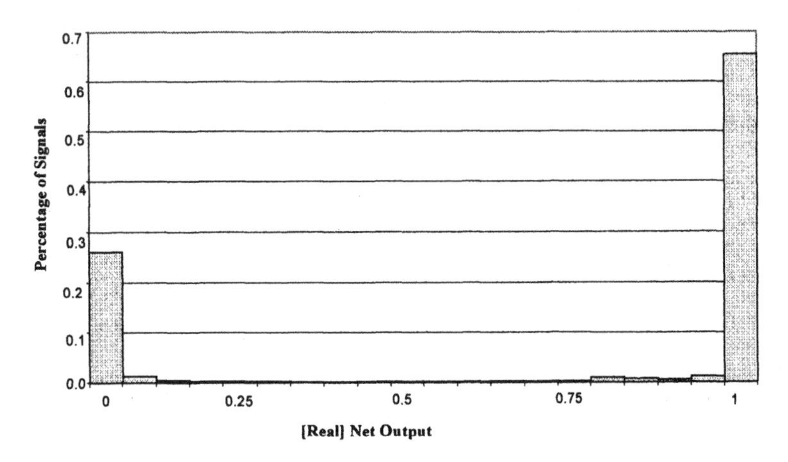

Fig. 9.7 Distribution of [real] net output for output variable NUTZPKW

Out of the total of 6,207 car users, the probit model classified 678 [10.9 percent] car users as non car users. This classification error is 10.9 percent compared with 1.5 percent for the CNN. Even more dramatic is the classification error of almost 50 percent [49.5 percent] for non car users. Non car users in particular were far better classified by the neural network. In total, the number of correctly identified observations with the probit model was 76.9 percent compared to 95 percent with the neural network.

9.7 Conclusions and Outlook

The application of backpropagation neural networks with a modified gradient descent learning rule in combination with different pruning procedures have been tested on the database of the German Mobility Panel. This panel was set up in 1994 by the German Ministry of Transport to collect data on mobility behaviour of individuals. The data base comprises some 10,000 individual observations. In our pilot study, methods from econometric modelling and neural networks were tested and compared for a continuous variable LOG[DAU_SUM], measuring the average daily time spent on mobility by an individual [in minutes] and a binary variable NUTZPKW, indicating the use of a car on a specific day.

The analysis of the variable DAU_SUM showed that the results of the panel model and neural network were comparable, with slightly better performance by the neural network. The variable 'NUTZPKW' was analysed using a probit model and also a backpropagation neural network with a modified gradient descent learning rule in combination with different pruning procedures. Here, the neural network provided clearly better out-of-sample performance.

These results demonstrate the effectiveness of the neural network approach, and should encourage their application to real world problems. However, it must be stressed that data needs to be carefully pre-processed and the network architecture selected via an appropriate search process. The splitting of the data set into training data, validation data and test data to void over fitting problems and to select networks with a high generalization ability also require considerable research experience. This suggests that there are strong barrier effects to the broad application of neural networks in social sciences. The development of self-generating neural networks which are much easier to apply is therefore of crucial importance and is one of the subjects of a current research programme.

References

Abuelgasim A.A., Gopal, S. and Strahler A.H. (1998): Forward and Inverse Modelling of Canopy Directional Reflectance Using a Neural Network, *International Journal of Remote Sensing* 19 (3), 1670-1672

Ackley D.H. and Littman M.S. (1994): A Case for Lamarckian Evolution. In: Langton C.G. (ed.) *Artificial Life III* [=SFI Studies in the Sciences of Complexity, Volume 18], Addison-Wesley, Reading [MA], pp. 3-10

Akaike H. (1974): A New Look at the Statistical Model Identification, *IEEE Transactions on Automatic Control* AC-19 (6), 716-723

Amari S. (1990): Mathematical Foundations of Neurocomputing, *Proceedings of the IEEE'78*, IEEE Press, Piscataway [NJ], pp. 1443-1463

Amari S. (1995): Information Geometry of the EM and EM Algorithms for Neural Networks, *Neural Networks* 8 (9), 1379-1409

Angeline P.J., Sauders G.M. and Pollack J.B. (1994): An Evolutionary Algorithm that Constructs Recurrent Neural Networks, *IEEE Transactions on Neural Networks* 5 (1), 54-65

Arbib M.A. (ed.) (1995): *The Handbook of Brain Theory and Neural Networks*, MIT Press, Cambridge [MA]

Argialas D.P. and Harlow C.A. (1990): Computational Image Processing Models: An Overview and Perspective, *Photogrammetric Engineering and Remote Sensing* 56, 871-886

Augusteijn M.F. and Warrender C.E. (1998): Wetland Classification Using Optical and Radar Data and Neural Network Classification, *International Journal of Remote Sensing* 19 (8), 1545-1560

Bäck T., Fogel D. and Michalewicz Z. (eds.) (1997): *Handbook of Evolutionary Computation*, Oxford University Press, New York

Baldi P. and Chauvin Y. (1991): Temporal Evolution of Generalization During Learning in Linear Networks, *Neural Computation* 3, 589-603

Baldi P. and Hornik K. (1989): Neural Networks and Principle Component Analysis. Learning from Examples without Local Minima, *Neural Networks* 2, 53-58

Baldi P. and Hornik K. (1995): Learning in Linear Neural Networks: A Survey, *IEEE Transactions on Neural Networks* 6 (4), 837-858

Baltagi B.H. (1995): *Econometric Analysis of Panel Data*, John Wiley, New York

Bankert R.L. (1994): Cloud Classification of AVHRR Imagery in Maritime Regions Using a Probabilistic Neural Network, *Journal of Applied Meteorology* 33, 322-333

Banks R.B. (1994): *Growth and Diffusion Phenomena: Mathematical Frameworks and Applications*, Springer, Berlin, Heidelberg, New York

Baraldi A. and Parmiggiani F. (1995): A Neural Network for Unsupervised Categorization of Multivalued Input Patterns: An Application to Satellite Image Clustering, *IEEE Transactions on Geoscience and Remote Sensing* 33, 305-316

Barnsley M. (1993): Monitoring Urban Areas in the EC Using Satellite Remote Sensing, *GIS Europe* 2 (8), 42-44

Batten D.F and Boyce D.E. (1986): Spatial Interaction, Transportation, and Interregional Commodity Flow Models. In: Nijkamp P (ed.) *Handbook of Regional and Urban Economics. Volume 1. Regional Economics*, North-Holland, Amsterdam, pp. 357-406

Battiti R. and Tecchiolli G. (1994): Learning with First, Second and No Derivatives: A Case Study in High Energy Physics, *Neurocomputing* 6, 181-206

Batty D. (1999): The Dynamics of Strike and Protest: France, May-June 1968, Unpublished Masters Dissertation, Department of Economic History, London School of Economics, London

Batty M. and Longley P. (1994): *Fractal Cities: A Geometry of Form and Function*, Academic Press, San Diego [CA]

Batty M., Xie Y. and Sun Z. (1999): Modelling Urban Dynamics through GIS-Based Cellular Automata, *Computers, Environment and Urban Systems* 23, 205-233

Baumann J.H, Fischer M.M. and Schubert U. (1983): A Multiregional Labour Supply Model for Austria: The Effects of Different Regionalizations in Multiregional Labour Market Modelling, *Papers of the Regional Science Association* 52, 53-83

Belew R.K., McInerney J. and Schraudolph N.N. (1991): Evolving Networks: Using Genetic Algorithm with Connectionist Learning, Technical Report No. CS90-174 (Revised), Computer Science & Engineering Department (C-014), University of California at San Diego [CA]

Benediktsson J.A. and Kanellopoulos I. (1999): Classification of Multisource and Hyperspectral Data Based on Decision Fusion, *IEEE Transactions on Geoscience and Remote Sensing*, 37 (3), 1367-1377

Benediktsson J.A., Swain P.H. and Ersoy O.K. (1990): Neural Network Approaches versus Statistical Methods in Classification of Multisource Remote Sensing Data, *IEEE Transactions on Geoscience and Remote Sensing* 28, 540-552

Benediktsson J.A., Swain P.H. and Ersoy O.K. (1993): Conjugate-Gradient Neural Networks in Classification of Multisource and Very-High-Dimensional Remote Sensing Data, *International Journal of Remote Sensing* 14 (15), 2883-2903

Benediktsson J.A., Sveinsson J.R., Ersoy O.K., and Swain P.H. (1997): Parallel Consensual Neural Networks, *IEEE Transactions on Neural Networks* 8 (1), 54-64

Bennett K.P. and Mangasarian O.L. (1992): Robust Linear Programming Discrimination of Two Linearly Inseparable Sets, *Optimization Methods and Software* 1, 23-34

Bernard A.C., Wilkinson G.G. and Kanellopoulos I. (1997): Training Strategies for Neural Network Soft Classification of Remotely Sensed Imagery, *International Journal of Remote Sensing* 18 (8), 1851-1856

Bezdek J.C. (1994): What's Computational Intelligence? In: Zurada J.M., Marks II R.J. and Robinson C.J. (eds.) *Computational Intelligence: Imitating Life*, IEEE Press, Piscataway [NJ], pp. 1-12

Bezdek J.C. and Pal S.K. (eds.) (1992): *Fuzzy Models for Pattern Recognition*, IEEE Press, Piscataway [NJ]

Bischof H., Schneider W. and Pinz A.J. (1992): Multispectral Classification of Landsat-Images Using Neural Networks, *IEEE Transactions on Geoscience and Remote Sensing* 30 (3), 482-490

Bishop C.M. (1995): *Neural Networks for Pattern Recognition*, Clarendon Press, Oxford

Blamire P.A. and Mineter M.J. (1995): On the Identification of Urban Areas from Landsat TM Data Using Artificial Neural Networks. In: Curran P.J. and Robertson Y.C. (eds.) *Proceedings of the 21rst Annual Conference of the Remote Sensing Society*, Southampton, Remote Sensing Society, Nottingham, pp. 50-57

Blumenfeld H. (1954): The Tidal Wave of Metropolitan Expansion, *Journal of the American Institute of Planners* 20, 3-14

Booker J.B., Goldberg D.E. and Holland J.H. (1989): Classifier Systems and Genetic Algorithms, *Artificial Intelligence* 40, 235-282

Bracken A.J. and Tuckwell H.C. (1992): Simple Mathematical Models for Urban Growth. *Proceedings of the Royal Society of London* A 438, pp. 171-181

Broomhead D.S. and Lowe D. (1988): Multivariable Functional Interpolation and Adaptive Networks, *Complex Systems* 2, 321-355

Bruzzone I., Conese C., Maselli F. and Roli F. (1997): Multisource Classification of Complex Rural Areas by Statistical and Neural-Network Approaches, *Photogrammetric Engineering and Remote Sensing* 63 (5), 523-533

Bruzzone L., Prieto D.F. and Serpico S.B. (1999): A Neural-Statistical Approach to Multitemporal and Multisource Remote-Sensing Image Classification, *IEEE Transactions on Geoscience and Remote Sensing* 37, 1350-1359

Buckton D., O'Mongain E. and Danaher S. (1999): The Use of Neural Networks for the Estimation of Oceanic Constituents Based on the MERIS Instrument, *International Journal of Remote Sensing* 20 (9), 1841-1851

Burgess N. (1994): A Constructive Algorithm that Converges for Real-Valued Input Patterns, *International Journal of Neural Systems* 5 (1), 59-66

Bussiere R. (ed.) (1972): Modeles Mathematiques de Repartition des Populations Urbaines, Centre de Recherche d'Urbanisme, Paris, France

Campbell W.J., Hill S.E. and Cromp R.F. (1989): Automatic Labelling and Characterization of Objects Using Artificial Neural Networks, *Telematics and Informatics* 6, 259-271

Cao Z., Kandel A. and Li L. (1990): A New Model of Fuzzy Reasoning, *Fuzzy Sets and Systems* 36, 311-325

Carpenter G.A. (1989): Neural Network Models for Pattern Recognition and Associative Memory, *Neural Networks* 2, 243-257

Carpenter G.A. and Grossberg S. (1985): Category Learning and Adaptive Pattern Recognition, A Neural Network Model, *Proceedings of the Third Army Conference on Applied Mathematics and Computing*, ARO-Report 86-1, pp. 37-56

Carpenter G.A. and Grossberg S. (1987a): A Massively Parallel Architecture for a Self-Organizing Neural Pattern Recognition Machine, *Computer Vision, Graphics, and Image Processing* 37, 54-115

Carpenter G.A. and Grossberg S. (1987b): ART 2 Stable Self-Organizing of Pattern Recognition Codes for Analog Input Patterns, *Applied Optics* 26, 4919-4930

Carpenter G.A. and Grossberg S. (eds.) (1991): *Pattern Recognition by Self-Organizing Neural Networks*, MIT Press, Cambridge [MA]

Carpenter G.A. and Grossberg S. (1995): Fuzzy ART. In: Kosko B. (ed.) *Fuzzy Engineering*, Prentice Hall, Carmel, pp. 467-497

Carpenter G.A., Grossberg S. and Reynolds J.H. (1991a) ARTMAP Supervised Real-Time Learning and Classification of Nonstationary Data by a Self-Organizing Neural Network, *Neural Networks* 4, 565-588

Carpenter G.A., Grossberg S. and Rosen D.B. (1991b): Fuzzy ART Fast Stable Learning and Categorization of Analog Patterns by an Adaptive Resonance System, *Neural Networks* 4, 759-771

Carpenter G.A., Grossberg S. and Ross W.D. (1993): ART-EMAP: A Neural Network Architecture for Object Recognition by Evidence Accumulation, *Proceedings of the World Congress on Neural Networks* (WCNN-93), Lawrence Ertbaum Associates, III, Hillsdak [NJ], pp. 643-656

Carpenter, G.A., Gjaja, M.N., Gopal, S. and Woodcock, C.E. (1997): ART Neural Networks for Remote Sensing: Vegetation Classification from Landsat TM and Terrain Data, *IEEE Transactions on Geoscience and Remote Sensing* 35, 308-325

Carpenter G.A., Grossberg S., Markuzon N., Reynolds J.H. and Rosen D.B. (1992): Fuzzy ARTMAP A Neural Network Architecture for Incremental Supervised Learning of Analog Multidimensional Maps, *IEEE Transactions on Neural Networks* 3, 698-713

Chen K.S., Tzeng Y.C., Chen C.F. and Kao W.L. (1995): Land-Cover Classification of Multispectral Imagery Using a Dynamic Learning Neural Network, *Photogrammetry Engineering and Remote Sensing* 61, 403-408

Chlond B., Lipps O. and Zumkeller D. (1996): Auswertung der Paneluntersuchung zum Verkehrsverhalten, Schlussbericht des Forschungsauftrags BMV FE 90442/95, Karlsruhe

Cichocki A. and Unbehauen R. (1993): *Neural Networks for Optimization and Signal Processing*, Wiley, Chichester

Civco D.L. (1993): Artificial Neural Networks for Land-Cover Classification and Mapping, *International Journal of Geographical Information Systems* 7, 173-186

Clark C. and Boyce J. (1999): The Detection of Ship Trail Clouds by Artificial Neural Network, *International Journal of Remote Sensing* 20 (4), 711-726

Clarke K.C. and Gaydos L.J. (1998): Loose-Coupling of a Cellular Automaton Model and GIS: Long-term Growth Prediction for the San Francisco and Washington/Baltimore Regions, *International Journal of Geographical Information Science* 12, 699-714

Cliff A.D., Haggett P. and Smallman-Raynor M. (1998): *Deciphering Global Epidemics: Analytical Approaches to the Disease Records of World Cities, 1888-1912,* Cambridge University Press, Cambridge

Cliff A.D., Haggett P., Ord J.K. and Versey G.R. (1981): *Spatial Diffusion: An Historical Geography of Epidemics in an Island Community*, Cambridge University Press, Cambridge

Congalton R.G. (1991): A Review of Assessing the Accuracy of Classifications of Remotely Sensed Data, *Remote Sensing Environment* 37, 35-46

Côté S. and Tatnall A.R.L. (1997): The Hopfield Neural Network as a Tool for Feature Tracking and Recognition from Satellite Sensor Images, *International Journal of Remote Sensing* 18, 871-885

Crowder R.S. (1990): Predicting the Mackey-Glass Timeseries with Cascade-Correlation Learning. In: Touretzky D S., Hinton G.E. and Sejnowski T.J. (eds) *Proceedings of the 1990 Connectionist Models Summer School*, Carnegie Mellon University, pp. 117-123

Darwen P. and Yao X. (1995): A Dilemma for Fitness Sharing with a Scaling Function, *Proceedings of the IEEE International Conference on Evolutionary Computation* [ICEC'95], Perth, Australia, IEEE Press, Piscataway [NJ], pp. 166-171

Darwen P. and Yao X. (1996a): Automatic Modularization by Specification, *Proceedings of the IEEE International Conference on Evolutionary Computation* [ICEC'96], Nagoya, Japan, IEEE Press, Piscataway [NJ], pp. 88-93

Darwen P. and Yao X. (1996b): Every Niching Method Has its Niche:Fitness Sharing and Implicit Sharing Compared. In: Voigt H.-M., Ebeling W., Rechenberg I. and Schwefel H.-P. (eds.) *Parallel Problem Solving from Nature* [PPSN] IV, Springer, Berlin, Heidelberg, New York, pp. 398-407

Darwen P. and Yao X. (1997): Speciation as Automatic Categorical Modularization, *IEEE Transactions on Evolutionary Computation* 1 (2), 101-108

Davis L. (ed.) (1991): *Handbook of Genetic Algorithms*, Van Nostrand Reinhold, New York

Dawson M.S., Fung A.K. and Manry M.T. (1993): Surface Parameter Retrieval Using Fast Learning Neural Networks, *Remote Sensing Reviews* 7, 1-18

De Jong K. and Spears W. (1992): A Formal Analysis of the Role of Multi-Point Crossover in Genetic Algorithms, *Annals of Mathematics and Artificial Intelligence* 5, 1-26

Dendrinos D.S. and Mullaly H. (1985): *Urban Evolution: Studies in the Mathematical Ecology of Cities*, Oxford University Press, Oxford

Di Carlo W. (2000): Exploring Multi-Dimensional Remote Sensing Data with a Virtual Reality System, *Geographical and Environmental Modelling* 4 (1), 7-20

Di Carlo W., and Wilkinson G.G. (1996): Dominant Linear Feature Detection in Satellite Images Using a Self-Organizing Neural Network. In: Binaghi E., Brivio P.A. and Rampini A. (eds.) *Soft Computing in Remote Sensing Data Analysis*, World Scientific, Singapore, pp. 73-79

Dony R.D. and Haykin S. (1995): Neural Network Approaches to Image Compression, *Proceedings of the IEEE International Conference on Evolutionary Computation* [ICEC'93], IEEE Press, Piscataway [NJ], [Volume 2], pp. 288-303

Downey I.D., Power C.H., Kanellopoulos I. and Wilkinson G.G. (1992): A Performance Comparison of Landsat TM Land Cover Classification Based on Neural Network Techniques and Traditional Maximum Likelihood and Minimum Distance Algorithms. In: Cracknell A.P. and Vaughan R.A. (eds.) *Proceedings of the 18th Annual Conference of the UK Remote Sensing Society*, Dundee, Scotland, Remote Sensing Society, Nottingham, pp. 518-528

Dreyer P. (1993): Classification of Land Cover Using Optimized Neural Nets on SPOT Data, *Photogrammetric Engineering and Remote Sensing* 59 (5), 617-621

Durrett R. (1995): Spatial Epidemic Models. In: Mollison D. (ed.) *Epidemic Models: Their Structure and Relation to Data*, Cambridge University Press, Cambridge, pp. 187-201

Efron B. and Tibshirani R. (1993): *An Introduction to the Bootstrap*, Chapman and Hall, New York

Eichmann G.E. and Stojancic M. (1987): Superresolving Signal and Image Restoration Using a Linear Associative Memory, *Applied Optics* 26, 1911-1918

Ellis J., Warner M. and White R.G. (1994): Smoothing SAR Images with Neural Networks, *Proceedings of the International Geoscience and Remote Sensing Symposium* [IGARSS'94], Pasadena, California, [Volume 4], IEEE Press, Piscataway [NJ], pp. 1883-1885

Epstein J.M. (1997): *Non-Linear Dynamics, Mathematical Biology, and Social Science*, Addison-Wesley, Reading [MA]

Fahlmann S.E. (1988): Faster-Learning Variations on Back-Propagation: An Empirical Study. In: Touretzky D., Hinton G. and Sejnowski T. (eds.) *Proceedings of the 1988 Connectionist Models Summer School*, Morgan Kaufmann Publishers, San Mateo [CA], pp. 38-51

Fahlman S.E. and Lebiere C. (1990): The Cascade-Correlation Learning Architecture. In: Touretzky D.S. (ed.) *Advances in Neural Information Processing Systems* 2, Morgan Kaufmann Publishers, San Mateo [CA], pp. 524-532

Farmer J.D. and Sidorowich J.J. (1987): Predicting Chaotic Time Series, *Physical Review Letters* 59 (8), 845-847

Fierens F., Kanellopoulos I., Wilkinson G.G. and Megier J. (1994): Comparison and Visualization of Feature Space Behaviour of Statistical and Neural Classifiers of Satellite Imagery, *Proceedings of the International Geoscience and Remote Sensing Symposium* [IGARSS'94], Pasadena, California, 8-12 August, [Volume 4], IEEE Press, Piscataway [NJ], pp. 1880-1882

Finnoff W. (1991): Complexity Measures for Classes of Neural Networks with Variable Weight Bounds, *Proceedings of the International Joint Conference on Neural Networks* [IJCNN'91], IEEE Press, Singapore, pp. 2624-2630

Finnoff W., Hergert F. and Zimmermann H.-G. (1993): Improving Model Selection by Non-Convergent Methods, *Neural Networks* 6, 771-783

Fischer M.M. (1994): Expert Systems and Artificial Neural Network for Spatial Analysis and Modelling – Essential Components for Knowledge-Based Geographical Information Systems, *Geographical Systems* 1, 221-235

Fischer M.M. (1995): Fundamentals in Neurocomputing. In: Fischer M.M., Sikos T. and Bassa L. (eds.) *Recent Developments in Spatial Information, Modelling and Processing*, Geomarket, Budapest, pp. 31-41

Fischer M.M. (1998a): Computational Neural Networks: An Attractive Class of Mathematical Models for Transportation Research. In: Himanen V., Nijkamp P. and Reggiani A. (eds.) *Neural Networks in Transport Applications*, Ashgate, Aldershot, pp. 3-20

Fischer M.M. (1998b): Computational Neural Networks - A New Paradigm for Spatial Analysis, *Environment and Planning A* 30(10), 1873-1892

Fischer M.M. (1999a): Intelligent GI Analysis. In: Stillwell J., Geertman S. and Openshaw S. (eds.) *Geographical Information and Planning*, Springer, Berlin, Heidelberg, New York, pp. 349-368

Fischer M.M. (1999b): Spatial Analysis: Retrospect and Prospect. In: Longley P., Goodchild M.F, Maguire P. and Rhind D.W. (eds.) *Geographical Information Systems: Principles, Technical Issues, Management Issues and Applications*, Wiley, New York, pp. 283-292

Fischer M.M. (2000): Methodological Challenges in Neural Spatial Interaction Modelling: The Issue of Model Selection. In: Reggiani A. (ed.) *Spatial Economic Science: New Frontiers in Theory and Methodology*, Springer, Berlin, Heidelberg, New York, pp. 89-101

Fischer M.M. (2001): Spatial Analysis. In: Smelser N.J. and Baltes P.B. (eds.) *International Encyclopedia of the Social & Behavioural Sciences*, Pergamon, Amsterdam [in press]

Fischer M.M. and Abrahart B. (2000): Neurocomputing - Tools for Geographers. In: Openshaw S., Abrahart B. and Harris T. (eds.) *GeoComputation*, Taylor & Francis, London, New York, pp. 187-217

Fischer M.M. and Getis A. (eds.) (1997): *Recent Developments in Spatial Analysis – Spatial Statistics, Behavioural Modelling, and Computational Intelligence*, Springer, Berlin, Heidelberg, New York

Fischer M.M. and Gopal S. (1993): Neurocomputing - A New Paradigm for Geographic Information Processing, *Environment and Planning A* 25, 757-760

Fischer M.M. and Gopal S. (1994a): Artificial Neural Networks. A New Approach to Modelling Interregional Telecommunication Flows, *Journal of Regional Science* 34, 503-527

Fischer M.M. and Gopal S. (1994b): Neurocomputing and Spatial Information Processing, *Proceedings of the Eurostat/DOSES Workshop on 'New Tools for Spatial Data Analysis'*, Lisbon, Eurostat, Luxembourg, pp. 55-68

Fischer M.M. and Gopal S. (1996): Spectral Pattern Recognition and Fuzzy ARTMAP: Design Features, System Dynamics and Real World Simulations, *Proceedings of the Fourth European Congress on Intelligent Technologies and Soft Computing* [EUFIT'96], Elite Foundation, Aachen, pp. 1664-1668

Fischer M.M. and Leung Y. (1998): A Genetic-Algorithm Based Evolutionary Computational Neural Network for Modelling Spatial Interaction Data, *The Annals of Regional Science* 32 (3), 437-458

Fischer M.M. and Reismann M. (2000): Evaluating Neural Spatial Interaction Modelling by Bootstrapping, Paper presented at the *6th World Congress of the Regional Science Association International*, Lugano, May 16-20, 2000 [to be published in *Network and Spatial Economics* 1 (2001)]

Fischer M.M. and Staufer P. (1999): Optimization in an Error Backpropagation Neural Network Environment with a Performance Test on a Spectral Pattern Classification Problem, *Geographical Analysis* 31(2), 89-108

Fischer M.M., Gopal S., Staufer P. and Steinnocher K. (1994): Evaluation of Neural Pattern Classifiers for a Remote Sensing Application. Paper presented at the *34th European Congress of the Regional Science Association*, Groningen, August 1994 [published 1997 in *Geographical Systems* 4 (2), 195-223]

Fischer M.M., Hlaváčková-Schindler K. and Reismann M. (1999): A Global Search Procedure for Parameter Estimation in Neural Spatial Interaction Modelling, *Papers in Regional Science* 78, 119-134

Fletcher R. (1986): *Practical Methods of Optimization*, Wiley, New York

Fogel, D.B. (1993). Using Evolutionary Programming to Create Neural Networks that are Capable of Playing Tic-Tac-Toe, *Proceedings of the 1993 International Joint Conference on Neural Networks* [IJCNN'93], IEEE Press, Piscataway [NJ], pp. 875-880

Fogel D.B. (1995a): *Evolutionary Computation: Towards a New Philosophy of Machine Intelligence*, IEEE Press, Piscataway [NJ]

Fogel D.B. (1995b): Phenotypes, Genotypes, and Operators in Evolutionary Computation, *Proceedings of the IEEE International Conference on Evolutionary Computation* [ICEC'95], Perth, Australia, IEEE Press, Piscataway [NJ], pp. 193-198

Fogel G.B. and Fogel D.B. (1995): Continuous Evolutionary Programming, *Analysis and Experiments, Cybernetics and Systems* 26, 79-90

Fogel D.B., Fogel L.J. and Porto V.W. (1990): Evolving Neural Networks, *Biological Cybernetics* 63, 487-493

Foody G.M. (1996): Relating the Land-Cover Composition of Mixed Pixels to Artificial Neural Network Classification Output, *Photogrammetric Engineering and Remote Sensing* 62 (5), 491-499

Foody G.M. and Boyd D.S. (1999): Fuzzy Mapping of Tropical Land Cover Along an Environmental Gradient from Remotely Sensed Data with an Artificial Neural Network, *Journal of Geographical Systems* 1, 23-35

Foody G.M., McCulloch M.B. and William B.Y. (1995): Classification of Remotely Sensed Data by an Artificial Neural Network: Issues Related to Training Data Characteristics, *Photogrammetric Engineering and Remote Sensing* 61, 391-401

Foody G.M., Lucas R.M., Curran P.J. and Honzak M. (1997): Non-Linear Mixture Modelling without End-Members Using an Artificial Neural Network, *International Journal of Remote Sensing* 18 (4), 937-953

Fotheringham A.S. (1983): A New Set of Spatial Interaction Models: The Theory of Competing Destinations, *Environment and Planning A* 15, 15-36

Fotheringham A.S. and O'Kelly M.E. (1989): *Spatial Interaction Models: Formulations and Applications*, Kluwer, Dordrecht, Boston

Frankhauser P. (1994): *La Fractalité des Structures Urbaine*, Collections Villes, Anthropos, Paris

Friedman J.H. and Stuetzle W. (1981): Projection Pursuit Regression, *Journal of the American Statistical Association* 76, 817-823

Gallant S.I. (1993): *Neural Network Learning and Expert Systems*, MIT Press, Cambridge [MA]

Goldberg D.E. (1989): *Genetic Algorithms in Search, Optimization and Machine Learning*. Addison-Wesley, Reading [MA]

Gong P., Pu R. and Chen J. (1996): Mapping Ecological Land Systems and Classification Uncertainties From Digital Elevation and Forest-Cover Data Using Neural Networks, *Photogrammetric Engineering and Remote Sensing* 62 (11), 1249-1260

Gong P. (1996): Integrated Analysis of Spatial Data From Multiple Sources: Using Evidential Reasoning and Artificial Neural Network Techniques For Geological Mapping, *Photogrammetric Engineering and Remote Sensing* 62 (5), 513-523

Gong P., Wang D.X. and Liang S. (1999): Inverting A Canopy Reflectance Model Using a Neural Network, *International Journal of Remote Sensing* 20 (1), 111-122

Gopal S. and Fischer M.M. (1993): Neural Net Based Interregional Telephone Traffic Models, *Proceedings of the International Joint Conference on Neural Networks* [IJCNN'93], Nagoya, Japan, pp. 2041-2044

Gopal S. and Fischer M.M. (1996): Learning in Single Hidden-Layer Feedforward Network Models, *Geographical Analysis* 28 (1), 38-55

Gopal S. and Fischer M.M. (1997): Fuzzy ARTMAP - A Neural Classifier For Multispectral Image Classification. In: Fischer M.M. and Getis A. (eds.) *Recent Developments in Spatial Analysis - Spatial Statistics, Behavioural Modelling and Computational Intelligence,* Springer, Berlin, Heidelberg, New York, pp. 306-335

Gopal S. and Woodcock C. (1996): Remote Sensing of Forest Change Using Artificial Neural Networks, *IEEE Transactions on Geoscience and Remote Sensing* 34, 398-403

Gopal S., Sklarew D.M. and Lambin E. (1993): Fuzzy Neural Network Classification of Landcover Change in the Sahel, *Proceedings of the Eurostat/DOSES Workshop on 'New Tools for Spatial Data Analysis'*, Lisbon, Eurostat, Luxembourg, pp. 69-81

Grossberg S. (1969): Some Networks that Can Learn, Remember, and Reproduce Any Number of Complicated Space-Time Patterns, *Journal of Mathematics and Mechanics* 19, 53-91

Grossberg S. (1976a): Adaptive Pattern Classification and Universal Recoding, I: Parallel Development and Coding of Neural Feature Detectors, *Biological Cybernetics* 23, 121-134

Grossberg S. (1976b): Adaptive Pattern Classification and Universal Recoding, II: Feedback, Expectation, Olfaction and Illusion, *Biological Cybernetics* 23, 187-202

Grossberg S. (1988): Non-Linear Neural Networks Principles, Mechanisms, and Architectures, *Neural Networks* 1, 17-61

Haining R.P. (1994): Designing Spatial Data Analysis Modules for Geographical Information Systems. In: Fotheringham S. and Rogerson, P. (eds.) *Spatial Analysis and GIS*, Taylor & Francis, London, pp. 45-63

Hall P. (1988) *Cities of Tomorrow: An Intellectual History of Urban Planning and Design in the Twentieth Century*, Basil Blackwell, Oxford, UK

Hanson S.J. and Pratt L.J. (1989): Comparing Biases for Minimal Network Construction With Back-Propagation. In: Touretzky D.S. (ed.) *Advances in Neural Information Processing* 1, Morgan Kaufmann Publishers, San Mateo [CA], pp. 177-185

Hara Y., Atkins R.G., Yueh S.H., Shin R.T. and Kong J.A. (1994): Application of Neural Networks to Radar Image Classification, *IEEE Transactions on Geoscience and Remote Sensing* 32, 100-109

Haring S., Viergever M.A. and Kok J.N. (1994): Kohonen Networks For Multiscale Image Segmentation, *Image and Vision Computing* 12 (6), 339-344

Hartigan J. (1975): *Clustering Algorithms*, Wiley, New York

Hashem S. (1993): Optimal Linear Combinations of Neural Networks, PhD Thesis, School of Industrial Engineering, Purdue University

Hautzinger H., Bäumer M., Heidemann D., Haag G. and Stackelberg B. von (2000): Erprobung und Evaluierung von Modellen der Statistik und der Künstlichen Intelligenz als Instrumente zur Analyse des Mobilitätspanels, Der Bundesminister für Verkehr, Bau- und Wohnungswesen, Bonn

Haynes K.E. and Fotheringham S. (1984): *Gravity and Spatial Interaction Models*, Sage Publications [Scientific Geography Series 2], Beverly Hills, London and New Delhi

Heermann P.D. and Khazenie N. (1992): Classification of Multispectral Remote Sensing Data Using a Back-Propagation Neural Network, *IEEE Transactions on Geoscience and Remote Sensing* 30, 81-88

Hepner G.F., Logan T., Rittner N. and Bryant N. (1990): Artificial Neural Network Classification Using a Minimal Training Set: Comparison to Conventional Supervised Classification, *Photogrammetric Engineering and Remote Sensing* 56, 469-473

Hernandez J.V., Moore K. and Elphic R. (1995): Sensor Fusion and Non-Linear Prediction for Anomalous Event Detection, *Proceedings of the SPIE International Society for Optical Engineering* 2484, SPIE Press, Beelingham, Washington, pp. 102-112

Hertz J., Krogh A. and Palmer R.G. (1991): *Introduction to the Theory of Neural Computation*, Addison-Wesley, Redwood City [CA]

Hestenes M.R. and Stiefel E. (1952): Methods of Conjugate Gradients for Solving Linear Systems, *Journal of Research of the National Bureau of Standards* 49 (6), 409-436

Hinton G.E. (1987): Learning Translation Invariant Recognition in Massively Parallel Networks. In: Bakker J.W. de, Nijman A.J. and Treleaven P.C. (eds.) *Proceedings of the Conference on Parallel Architectures and Languages Europe* [PARLE], Springer, Berlin, Heidelberg, New York, pp. 1-13

Hlavácková-Schindler K. and Fischer M.M. (2000): An Incremental Algorithm for Parallel Training of the Size and the Weights in a Feedforward Neural Network, *Neural Processing Letters* 11 (2), 131-138

Holland J.H. (1975): *Adaptation in Natural and Artificial Systems*, University of Michigan, Ann Arbor

Holland J.H. (1986): Escaping Brittleness: The Possibilities of General Purpose Learning Algorithms Applied to Parallel Rule-Based Systems. In: Michalski R.S., Carbonell J.G. and Michell T.M. (eds.) *Machine Learning: An Artificial Intelligence Approach* [Volume 2], Morgan Kaufmann Publishers, San Mateo [CA], pp. 593-624

Holland J.H. and Reitman J.S. (1978): Cognitive Systems Based on Adaptive Algorithms. In: Wataman D.A. and Hayes-Roth F. (eds.) *Pattern-Directed Inference Systems*, Academic Press, New York, pp. 313-329

Hornik K., Stinchcombe M. and White H. (1989): Multilayer Feedforward Networks are Universal Approximators, *Neural Networks* 2, 359-366

Hruschka H. and Natter M. (1992): Using Neural Networks for Clustering-Based Market Segmentation, Research Memorandum 307, Institute for Advanced Studies, Vienna

Hsiao C. (1986): *Analysis of Panel Data,* Cambridge University Press, Cambridge

Huang C.F. and Leung Y. (1999): Estimating the Relationship Between Isoseismal Area and Earthquake Magnitude by a Hybrid Fuzzy-Neural-Network Method, *Fuzzy Sets and Systems* 107, 131-146

Hugo J. (1998): Untersuchung der Anwendbarkeit von neuronalen Netzwerken und von Neuro-Fuzzy-Systemen in den Sozialwissenschaften, Ph.D. Thesis, University of Stuttgart

Hush D.R. and Horne B.G. (1993) Progress in Supervised Neural Networks, *IEEE Signal Processing Magazine* 10 (1), 8-39

Ishibuchi H., Kozaki K. and Tanaka H. (1992): Distributed Representation of Fuzzy Rules and its Application to Pattern Classification, *Fuzzy Sets and Systems* 52, 21-32

Ishibuchi H., Tanaka H. and Okada H. (1993): Fuzzy Neural Networks With Fuzzy Weights and Fuzzy Biases, *Proceedings of the IEEE International Conference on Neural Networks* [ICNN'93], IEEE Press, Piscataway [NJ], pp. 1650-1655

Ishibuchi H., Nozaki K., Yamamoto N. and Tanaka H. (1995): Selecting Fuzzy If-Then Rules for Classification Problems Using Genetic Algorithms, *IEEE Transactions on Fuzzy Systems* 3, 260-270

Ito Y. and Omatu S. (1997): Category Classification Method Using a Self-Organizing Neural Network, *International Journal of Remote Sensing* 18 (4), 829-845

Jackson K.T. (1985): *Crabgrass Frontier: The Suburbanization of the United States*, Oxford University Press, New York

Jacobs R.A. (1988): Increased Rates of Convergence Through Learning Rate Adaptation, *Neural Networks* 1 (4), 295-307

Jimenez L., Morales-Morell A. and Creus A. (1999): Classification of Hyperdimensional Data Based on Feature and Decision Fusion Approaches Using Projection Pursuit, Majority Voting, and Neural Networks, *IEEE Transactions on Geoscience and Remote Sensing* 37 (3), 1360-1366

Jin Y.-Q. and Liu C. (1997): Biomass Retrieval from High-Dimensional Active/Passive Remote Sensing Data by Using Artificial Neural Networks, *International Journal of Remote Sensing* 18 (4), 971-979

Johansson E.M., Dowla F.U. and Goodman D.M. (1991): Backpropagation Learning for Multi-Layer Feed-Forward Neural Networks Using the Conjugate Gradient Method, *International Journal of Neural Systems* 2 (4), 291-301

Kanellopoulos I., Wilkinson G.G. and Mégier J. (1993): Integration of Neural Network and Statistical Image Classification for Land Cover Mapping, *Proceedings of the International Geoscience and Remote Sensing Symposium* [IGARSS'93], Tokyo, [Volume 2], IEEE Press, Piscataway [NJ], pp. 511-513

Kanellopoulos I., Varfis A., Wilkinson G.G. and Mégier J. (1992): Land-Cover Discrimination in SPOT HRV Imagery Using an Artificial Neural Network: A 20-Class Experiment, *International Journal of Remote Sensing* 13, 917-924

Kanellopoulos I., Wilkinson G.G., Roli F. and Austin J. (eds.) (1997): *Neurocomputation in Remote Sensing Data Analysis*, Springer, Berlin, Heidelberg, New York

Kavzoglu T. and Mather P.M. (1999): Pruning Artificial Neural Networks: An Example Using Land Cover Classification of Multi-Sensor Images, *International Journal of Remote Sensing* 20 (14), 2787-2803

Key J., Maslanik J.A. and Schweiger A.J. (1989): Classification of Merged AVHRR and SMMR Arctic Data With Neural Networks, *Photogrammetric Engineering and Remote Sensing* 55 (9), 1331-1338

Kim T.J., Wiggins L.L. and Wright J.R. (eds.) (1990): *Expert Systems: Applications to Urban and Regional Planning*, Kluwer, Dordrecht, Boston, pp. 191-201

Kimes D.S., Ranson K.J. and Sun G. (1997): Inversion of a Forest Backscatter Model Using Neural Networks, *International Journal of Remote Sensing* 18 (10), 2181-2199

Kimes D.S., Nelson R.F., Manry M.T. and Fung A.K. (1998): Attributes of Neural Networks for Extracting Continuous Vegetation Variables from Optical and Radar Measurements, *International Journal of Remote Sensing* 19, 2639-2663

Kohonen T. (1982): Self-Organized Formation of Topologically Correct Feature Maps, *Biological Cybernetics* 43, 59-69

Kohonen T. (1988): *Self-Organization and Associative Memory*. Springer, Berlin, Heidelberg, New York [1rst edition 1984]

Kosko B. (1988): Bidirectional Associative Memory, *IEEE Transactions on Systems, Man, and Cybernetics* 18, 49-60

Kosko B. (1992): *Neural Networks and Fuzzy Systems*, Prentice Hall, Englewood Cliffs

Krugman P.R. (1993) First Nature, Second Nature, and Metropolitan Growth, *Journal of Regional Science* 33, 129-144

Le Cun Y. (1989): Generalization and Network Design Strategies. In: Pfeifer M. (ed.) *Connections in Perspective*, North-Holland, Amsterdam, pp. 143-155

Le Cun Y., Denker J.S. and Solla S.A. (1990): Optimal Brain Damage. In: Touretzky D.S. (ed.) *Advances in Neural Information Processing* 2, Morgan Kaufmann Publishers, San Mateo [CA], pp. 598-605

Lee H.M. and Lu B.H. (1994): FUZZY BP: A Neural Network Model with Fuzzy Inference, *Proceedings of the IEEE International Conference on Neural Networks*, Orlando [FL], IEEE Press, Piscataway [NJ], pp. 1583-1588

Lee H.M., Lu B.H. and Lin F.T. (1995): A Fuzzy Neural Network Model for Revising Imperfect Fuzzy Rules, *Fuzzy Sets and Systems* 76, 25-45

Lee J., Weyer R.C., Sengupta S.K. and Welch R.M. (1990): A Neural Network Approach to Cloud Classification, *IEEE Transactions on Geoscience and Remote Sensing* 28, 846-855

Leung K.S., Leung Y., So L. and Yam K.F. (1992a): Rule Learning in Expert Systems Using Genetic Algorithm: 1, Concepts, *Proceedings of the Second International*

Conference on Fuzzy Logic and Neural Networks, Jizuka, Kyushu Institute of Technology, pp. 201-204

Leung K.S., Leung Y., So L. and Yam K.F. (1992b): Rule Learning in Expert Systems Using Genetic Algorithm: 2, Empirical Studies, *Proceedings of the Second International Conference on Fuzzy Logic and Neural Networks*, Jizuka, Kyushu Institute of Technology, pp. 205-208

Leung Y. (1993): Towards the Development of an Intelligent Support System. In: Fischer M.M. and Nijkamp P. (eds.) *Geographic Information Systems, Spatial Modelling, and Policy Evaluation*, Springer, Berlin, Heidelberg, New York, pp. 131-145

Leung Y. (1997a): *Intelligent Spatial Decision Support Systems*. Springer, Berlin, Heidelberg, New York

Leung Y. (1997b): Feedforward Neural Network Models for Spatial Data Pattern Classification. In: Fischer M.M. and Getis A. (eds.) *Recent Developments in Spatial Analysis - Spatial Statistics, Behavioural Modelling and Computational Intelligence*, Springer, Berlin, Heidelberg, New York, pp. 336-359

Leung Y. and Lin X. (1996): Fast Extraction of Fuzzy and Non-Fuzzy IF-THEN Rules By a Radial Basis Function Network with Unsupervised Competitive Learning, Department of Geography, Chinese University of Hongkong [Unpublished Paper]

Leung Y., Dong T.X. and Xu Z.B. (1998): The Optimal Encoding of Biased Association in Linear Associative Memories, *Neural Networks* 11, 877-884

Leung Y., Gao Y. and Zhang W.X. (1997): A Novel Genetic-Based Method for Training Fuzzy System, Department of Geography, Chinese University of Hongkong [Unpublished Paper]

Leung Y., Leung K.S., Ng W. and Lau M.I. (1995): Evolving Multilayer Feedforward Neural Networks by Genetic Algorithms, Department of Geography, Chinese University of Hongkong [Unpublished Paper]

Leung Y., Luo J.C. and Zhou C.H. (2000): A Knowledge-Integrated Radial Basis Function Model for the Classification of Remotely Sensed Images, Department of Geography, Chinese University of Hongkong [Unpublished Paper]

Leung Y., Zhang J.S. and Xu Z.B. (1997): Neural Networks for Convex Hull Computation, *IEEE Transactions on Neural Networks* 8, 601-611

Leung Y., Zhang J.S. and Xu Z.B. (2000): Clustering By Scale-Space Filtering, *IEEE Transactions on Pattern Analysis and Machine Intelligence* [in press]

Lewis H.G., Côté S. and Tatnall A.R.L. (1997): Determination of Spatial and Temporal Characteristics as an Aid to Neural Network Cloud Classification, *International Journal of Remote Sensing* 18, 899-915

Lin C.T. and George C.S. (1991): Neural-Network Based Fuzzy Logic Control and Decision Systems, *IEEE Transactions on Computers* 40 (12), 1320-1336

Liu Y. and Yao X. (1996a): Evolutionary Design of Artificial Neural Networks With Different Nodes, *Proceedings of the IEEE International Conference on Evolutionary Computation* [ICEC'96], Nagoya, Japan, IEEE Press, Piscataway [NJ], pp. 670-675

Liu Y. and Yao X. (1996b): A Population-Based Learning Algorithm which Learns Both Architectures and Weights of Neural Networks, *Chinese Journal of Advanced Software Research* 3 (1), 54-65

Liu Y. and Yao X. (1997): Negatively Correlated Neural Networks Can Produce Best Ensembles, *Australian Journal of Intelligent Information Processing Systems* 4 (3/4), 176-185

Liu Y. and Yao X. (1999a): Ensemble Learning Via Negative Correlation, *Neural Networks* 12, 1391-1398

Liu Y. and Yao X. (1999b): Simultaneous Training of Negatively Correlated Neural Networks in an Ensemble, *IEEE Transactions on Systems, Man, and Cybernetics*, Part B: Cybernetics, 29(6), 716-725

Logar A., Corwin E., Alexander J., Lloyd D., Berendes T. and Welch R. (1997): A Hybrid Histogram / Neural Network Classifier For Creating Global Cloud Masks, *International Journal of Remote Sensing* 18 (4), 847-869

Longley P.A., Brooks S.M., McDonnell R. and Macmillan B. (eds.) (1998): *Geocomputation - A Primer*, John Wiley, Chichester [UK]

Luenberger P. (1984): *Linear and Non-linear Programing*, Addison-Wesley, Reading [MA]

Mackey M. and Glass L. (1977). Oscillation and Chaos in Physiological Control Systems, *Science* 197, 287

Maniezzo V. (1994). Genetic Evolution of the Topology and Weight Distribution of Neural Networks, *IEEE Transactions on Neural Networks* 5 (1), 39-53

Mannan B., Roy J. and Ray A.K. (1998): Fuzzy ARTMAP Supervised Classification of Multi-Spectral Remotely-Sensed Images, *International Journal of Remote Sensing* 19, 767-777

Martinetz T.M., Berkovich S G. and Schulten K.J. (1993): 'Neural-Gas' Network for Vector Quantization and Its Application to Time-Series Prediction, *IEEE Transactions on Neural Networks* 4 (4), 558-569

McClellan G.E., DeWitt R.N., Hemmer T.H., Matheson L.N. and Moe G.O. (1989): Multispectral Image-Processing with a Three-Layer Backpropagation Network, *Proceedings of the International Joint Conference on Neural Networks* [IJCNN'89], Washington, pp. 151-153

McDonnell J.R. and Waagen D. (1993): Neural Network Structure Design by Evolutionary Programming. In: Fogel D.B. and Atmar W. (eds.) *Proceedings of the Second Annual Conference on Evolutionary Programming*, Evolutionary Programming Society, La Jolla [CA], pp. 79-89

McDonnell J.R. and Waagen D. (1994): Evolving Recurrent Perceptrons for Time-Series Modelling, *IEEE Transactions on Neural Networks* 5 (1), 24-38

Michie D., Spiegelhalter D.J. and Taylor C.C. (1994): *Machine Learning, Neural and Statistical Classification*, Ellis Horwood, London

Miller G.F., Todd P.M. and Hegde S.U. (1989): Designing Neural Networks Using Genetic Algorithms. In: Schaffer J.D. (ed.) *Proceedings of the Third International Conference on Genetic Algorithms and their Applications*, Morgan Kaufmann Publishers, San Mateo [CA], pp. 379-384

Moddy A., Gopal S. and Strahler A.H. (1996): Artificial Neural Network Response to Mixed Pixels in Coarse-Resolution Satellite Data, *Remote Sensing Environment* 58, 329-343

Møller M.F. (1993): A Scaled Conjugate Gradient Algorithm for Fast Supervised Learning, *Neural Networks* 6 (4), 525-533

Monostori L. and Egresits C. (1994): Modelling and Monitoring of Milling Through Neuro-Fuzzy Techniques, *Proceedings of Intelligent Manufacturing Systems* [IMS'94], Vienna, 463-468

Moody J. and Darken C.J. (1989): Fast Learning in Networks of Locally-Tuned Processing Units, *Neural Computation* 1, 281-294

Morrill R.L. (1968): Waves of Spatial Diffusion, *Journal of Regional Science* 8, 1-18

Morris A.J. (1979): *History of Urban Form: Before the Industrial Revolutions*, Longman, London

Mozer M.C. and Smolensky P. (1989): Skeletonization: A Technique for Trimming the Fat from a Network via Relevance Assessment. In: Touretzky D.S. (ed.) *Advances in Neural Information Processing Systems* 1, Morgan Kaufmann Publishers, San Mateo [CA], pp. 107-115

Mulgrew B. and Cowan C.F.N. (1988): *Adaptive Filters and Equalisers,* Kluwer, Dordrecht, Boston

Müller K. and Haag G. (1996): The Austrian Innovation System: Complex Modelling with NIS Data, Institute for Advanced Studies, Vienna

Murai H. and Omatu, S. (1997): Remote Sensing Image Analysis Using a Neural Network and Knowledge-Based Processing, *International Journal of Remote Sensing* 18, 811-828

Murray J.D. (1987): Modelling the Spread of Rabies, *American Scientist* 75, 280-284

Murray J.D. (1993): *Mathematical Biology* [2nd edition], Springer, Berlin, Heidelberg, New York

Nadal J.-P. (1989): Study of a Growth Algorithm for a Feedforward Network, *International Journal of Neural Systems* 1, 55-59

Nauck D. and Kruse R. (1993): A Fuzzy Neural Network Learning Fuzzy Control Rules and Membership Function by Fuzzy Error Backpropagation, *Proceedings of the IEEE International Joint Conference on Neural Networks* [IJCNN'93], Nagoya, Japan, IEEE Press, Piscataway [NJ], pp. 1022-1027

Nivola P.S. (1999): *Laws of the Landscape: How Policies Shape Cities in Europe and America,* Brookings Institution Press, Washington [DC]

Noble J.V. (1974): Geographic and Temporal Development of the Plague, *Nature* 250, 726-729

Odri S.V., Petrovacki D.P. and Krstonosic G.A. (1993): Evolutional Development of a Multilevel Neural Network, *Neural Networks* 6 (4), 583-595

Openshaw S. (1988): Building an Automated Modelling System to Explore a Universe of Spatial Interaction Models, *Geographical Analysis* 20 (1), 31-46

Openshaw S. (1993): Modelling Spatial Interaction Using a Neural Net. In: Fischer M.M. and Nijkamp P. (eds.) *Geographic Information Systems, Spatial Modelling, and Policy Evaluation*, Springer, Berlin, Geidelberg, New York, pp. 147-164

Openshaw S. (1995): Developing Automated and Smart Spatial Pattern Exploration Tools for Geographical Systems Applications, *The Statistician* 44 (1), 3-16

Openshaw S. and Abrahart R.J. (eds.) (2000): *GeoComputation*, Taylor & Francis, London, New York

Openshaw S. and Openshaw C. (1997): *Artificial Intelligence in Geography*, John Wiley, New York

Openshaw S. and Taylor P. (1979): A Million or so Correlation Coefficients: Three Experiments on the Modifiable Areal Unit Problem. In: Bennett R.J, Thrift N.J. and Wrigley, N. (eds.) *Statistical Applications in the Spatial Sciences*, Pion, London, pp. 127-144

Openshaw S. and Wymer S. (1995): Classifying and Regionalising Census Data. In: Openshaw, S. (ed.) *Census Users Handbook*, Geoinformation International, Cambridge, pp. 353-361

Openshaw S., Fischer M.M., Benwell G. and Macmillan B. (2000): GeoComputation Research Agendas and Futures. In Openshaw S. and Abrahart R.J. (eds.) *GeoComputation*, Taylor & Francis, London, New York, pp. 379-400

Pankiewicz G.S. (1997): Neural Network Classification of Convective Airmasses for a Flood Forecasting System, *International Journal of Remote Sensing* 18 (4), 887-898

Paola J.D. and Schowengerdt R.A. (1995): A Review and Analysis of Backpropagation Neural Networks for Classification of Remotely-Sensed Multi-Spectral Imagery, *International Journal of Remote Sensing* 16 (16), 3033-3058

Paola I.D. and Schowengerdt R.A. (1997): The Effect of Neural-Network Structure on a Multispectral Land-Use/Land-Cover Classification, *Photogrammetric Engineering and Remote Sensing* 63, 535-544

Park D., Kandel A. and Langholz G. (1994): Genetic-Based New Fuzzy Reasoning Models with Application to Fuzzy Control, *IEEE Transactions on Systems, Man, and Cybernetics* 24 (1), 39-47

Peddle D.R., Foody G.M., Zhang A., Franklin S.E. and Ledrew E.F. (1994): Multisource Image Classification II: An Empirical Comparison of Evidential Reasoning, Linear Discriminant Analysis, and Maximum Likelihood Algorithms for Alpine Land Cover Classification, *Canadian Journal of Remote Sensing* 20, 397-408

Perrone M.P. (1993): Improving Regression Estimation: Averaging Methods for Variance Reduction with Extensions to General Convex Measure Optimization, PhD Thesis, Department of Physics, Brown University

Petrakos M., Di Carlo W. and Kanellopoulos I. (1999): Dimensionality Reduction for the Visualisation of Fuzzy Multispectral Data, *Proceedings of the 25th Annual Conference of the UK Remote Sensing Society,* Cardiff, Remote Sensing Society, Nottingham, pp. 55-62

Pierce L.E., Sarabandi K. and Ulaby F.T. (1994): Application of an Artificial Neural Network in Canopy Scattering Inversion, *International Journal of Remote Sensing* 15 (16), 3263-3270

Poggio T. and Girosi F. (1989): *A Theory of Networks for Approximation and Learning,* A.I. Memo No. 1140, MIT Press, Cambridge

Powell M.J.D. (1992): Radial Basis Functions in 1990, *Advanced Numerical Analysis* 2, 105-210

Prechelt L. (1994): Proben1 – A Set of Neural Network Benchmark Problems and Benchmarking Rules, Technical Report 21/94, Informatics Faculty, University of Karlsruhe, Karlsruhe, Germany

Prechelt L. (1995): Some Notes on Neural Learning Algorithm Benchmarking, *Neurocomputing* 9 (3), 343-347

Press W.H., Teukolsky S.A., Vetterling W.T. and Flannery B.P. (1992): *Numerical Recipes in C. The Art of Scientific Computing,* Cambridge University Press, Cambridge

Raggett G.F. (1982): Modelling the Eyam Plague, *Bulletin of the Institute of Mathematics and its Applications* 18, 221-226

Reed R. (1993): Pruning Algorithms - A Survey, *IEEE Transactions on Neural Networks* 4 (5), 740-747

Rhee F.C.-H. and Krishnapuram R. (1993): Fuzzy Rule Generation Methods for High-Level Computer Vision, *Fuzzy Sets and Systems* 60, 245-258

Rissanen J. (1978): Modelling by Shortest Data Description, *Automatica* 14 (5), 465-471

Richardson L.F. (1941): Mathematical Theory of Population Movement, *Nature* 148, 357

Rollet R., Benie G.B., Li W. and Wang S. (1998): Image Classification Algorithm Based on the RBF Neural Network and K-Means, *International Journal of Remote Sensing* 19, 3003-3009

Roy A., Govil S. and Miranda R. (1995): An Algorithm to Generate Radial Basis Function (RBF)-Like Nets for Classification Problems, *Neural Networks* 8, 179-201

Rumelhart D.E., McClelland J.L. and the PDP Research Group (eds.) (1986): *Parallel Distributed Processing: Exploration in the Microstructure of Cognition, Volume I: Foundations,* MIT Press, Cambridge [MA]

Rumelhart D.E., Hinton G.E. and Williams R.J. (1986): Learning Internal Representations by Error Propagations. In: Rumelhart D.E., McClelland J.L. and the PDP Research Group (eds.) *Parallel Distributed Processing: Explorations in the Microstructures of Cognition, Volume 1: Foundations*. MIT Press, Cambridge [MA], pp. 318-362

Saratchandran P., Sundararajan N. and Foo S.K. (1996): *Parallel Implementations of Backpropagation Neural Networks on Transputers: A Study of Training Set Parallelism,* World Scientific, Singapore

Schaale M. and Furrer R. (1995): Land Surface Classification by Neural Networks, *International Journal of Remote Sensing* 16 (16), 3003-3031

Schaffer J.D. (1985): Multiple Objective Optimization with Vector Evaluated Genetic Algorithms, *Proceedings of the First International Conference on Genetic Algorithms and their Applications*, pp. 93-100

Schalkoff R.J. (1997): *Artificial Neural Networks*, McGraw-Hill, New York

Schiffmann W., Joost M. and Werner R. (1992): Synthesis and Performance Analysis of Multilayer Neural Network Architectures, Technical Report 16/1992, University of Koblenz, Institute of Physics, Koblenz

Schiffmann W., Joost M. and Werner R. (1993): Comparison of Optimized Backpropagation Algorithms, *Proceedings of the European Symposium on Artificial Neural Networks* [ESANN '93], Brussels, Belgium, pp. 97-104

Schweiger A.J. and Key J.R. (1997): Estimating Surface Radiation Fluxes in the Arctic From TOVS Brightness Temperatures, *International Journal of Remote Sensing* 18 (4), 955-970

Serpico S.B. and Roli F. (1995): Classification of Multisensor Remote-Sensing Images by Structured Neural Networks, *IEEE Transactions on Geoscience and Remote Sensing* 33 (3), 562-578

Setiono R. and Hui L.C.K. (1995): Use of a Quasi-Newton Method in a Feed-Forward Neural Network Construction Algorithm, *IEEE Transactions on Neural Networks* 6 (1), 273-277

Shanno D.F. (1990): Recent Advances in Numerical Techniques for Large-Scale Optimization. In: Miller W.T. (ed.) *Neural Networks for Robotics and Control*, MIT Press, Cambridge [MA], pp. 171-178

Short N. (1991): A Real-Time Expert System and Neural Network for the Classification of Remotely Sensed Data, *Proceedings of the Annual Convention on the American Society for Photogrammetry and Remote Sensing* 3, American Society for Photogrammetry and Remote Sensing, Pittsburg, pp. 406-418

Sietsma J. and Dow R.J.F. (1991): Creating Artificial Neural Networks that Generalize, *Neural Networks* 4, 67-79

Skidmore A.K., Turner B.J., Brinkhof W. and Knowles E. (1997): Performance of a Neural Network: Mapping Forest Using GIS and Remotely Sensed Data, *Photogrammetric Engineering and Remote Sensing* 63, 501-514

Smalz R. and Conrad M. (1994): Combining Evolution with Credit Apportionment: A New Learning Algorithm for Neural Nets, *Neural Networks* 7 (2), 341-351

Smith S.F. (1983): Flexible Learning of Problem Solving Heuristics through Adaptive Search, *Proceedings of the Eighth International Joint Conference of Artificial Intelligence*, pp. 422-425

Smith T.R., Penquet R., Menon S. and Agarwal P. (1987): KBGIS-II. A Knowledge-Based Geographical Information System, *International Journal of Geographic Information Systems* 1, 149-172

Spina M.S., Schwartz M.J., Staelin D.H. and Gasiewski A.J. (1994): Application of Multilayer Feedforward Neural Networks to Precipitation Cell-Top Altitude Estimation, *Proceedings of the International Geoscience and Remote Sensing Symposium* Pasadena [CA], [IIGARSS '94], [Volume 4], IEEE Press, Piscataway [NJ], pp. 1870-1872

Stackelberg B. von (1998): Entwicklung eines Vortrainingsverfahrens für neuronale Netze, MA Thesis, University of Stuttgart

Steele N.C., Reeves C.R., Nicholas M. and King P.J. (1995): Radial Basis Function Artificial Neural Networks for the Inference Process in Fuzzy Logic Based Control, *Computing* 54, 99-117

Stephanidis C.N., Cracknell A.P., Hayes L.W.B. and Slogett D.R. (1995): The Implementation of a Self-Organised Neural Network for Algal Bloom Detection. In: Curran P.J. and Robertson Y.C. (eds.) *Proceedings of the 24rst Annual Conference of the UK Remote Sensing Society*, Southampton, Remote Sensing Society, Nottingham, pp. 19-25

Stork D. and Allen J. (1992): How to Solve the N-bit Parity Problem with Two Hidden Units, *Neural Networks* 5 (6), 923-926

Storn R. and Price K. (1996): Minimizing the Real Functions of the ICEC'96 Contest by Differential Evolution, *Proceedings of the IEEE International Conference on Evolutionary Computation*, Nagoya, Japan, IEEE Press, Piscataway [NJ], pp. 842-844

Storn R. and Price K. (1997): Differential Evolution - A Simple and Efficient Heuristic for Global Optimization Over Continuous Spaces, *Journal of Global Optimization* 11, 341-359

Sulzberger S.M., Tschichold-Gürman N.N. and Vestli S.J. (1993): FUN: Optimization of Fuzzy Rule Based Systems Using Neural Networks, *Proceedings of the IEEE International Conference on Neural Networks*, IEEE Press, Piscataway [NJ], pp. 312-316

Syswerda G. (1989): Uniform Crossover in Genetic Algorithms, *Proceedings of the Third International Conference of Genetic Algorithms*, Morgan Kaufman, San Mateo [CA]

Syswerda G. (1991): A Study of Reproduction in Generational and Steady State Genetic Algorithms. In: Rawlins G.J.E. (ed.) *Foundations of Genetic Algorithms*, Morgan Kaufmann Publishers, San Mateo [CA], pp. 94-101

Thackrah G., Barnsley M. and Pan P. (1999): Merging Land Cover Classifications from Artificial Neural Networks Using the Sogeno Fuzzy Integral, *Proceedings of the 25th Annual Conference of the UK Remote Sensing Society*, Cardiff, Remote Sensing Society, Nottingham, pp. 71-78

Tibshirani R. (1996): A Comparison of Some Error Estimates for Neural Network Models, *Neural Computation* 8 (1), 152-163

Toffoli T. and Margolus N. (1987): *Cellular Automata Machines: A New Environment for Modelling*, MIT Press, Cambridge [MA]

Tso B.C.K. and Mather P.M. (1999): Classification of Multisource Remote Sensing Imagery Using a Genetic Algorithm and Markov Random Fields, *IEEE Transactions on Geoscience and Remote Sensing* 37, 1255-1260

Turton I., Openshaw S. and Diplock G.J. (1997): A Genetic Programming Approach to Building New Spatial Model Relevant to GIS. In: Kemp Z. (ed.) *Innovations in GIS 4*, Taylor & Francis, London, pp. 89-102

Tzeng Y.C. and Chen K.S. (1998): A Fuzzy Neural Network to SAR Image Classification, *IEEE Transactions on Geoscience and Remote Sensing* 36, 301-307

Tzeng Y.C., Chen K.S., Kao W.L. and Fung A.K. (1994): A Dynamic Learning Neural Network for Remote Sensing Applications, *IEEE Transactions on Geoscience and Remote Sensing* 32, 1096-1102

Walder P. and MacLaren I. (2000): Neural Network Based Methods for Cloud Classification on AVHRR Images, *International Journal of Remote Sensing* 21, 1693-1708

Wallace C.S. and Patrick J.D. (1991): Coding Decision Trees, Technical Report 91/153, Department of Computer Science, Monash University, Clayton, Victoria, Australia

Wan W. and Fraser D. (1999): Multisource Data Fusion With Multiple Self-Organizing Maps, *IEEE Transactions on Geoscience and Remote Sensing* 37, 1344-1351

Wang F. (1994): The Use of Artificial Neural Networks in a Geographical Information System for Agricultural Land-Suitability Assessment, *Environment and Planning* A 26, 265-284

Wang Y. and Dong D. (1997): Retrieving Forest Stand Parameters from SAR Backscatter Data Using a Neural Network Trained by a Canopy Backscatter Model, *International Journal of Remote Sensing* 18 (4), 981-989

Wann C.-D. and Thomopoulos S.C.A. (1997): Application of Self-Organizing Neural Networks to Multiradar Data Fusion, *Optical Engineering* 36 (3), 799-813

Ward B. and Redfern S. (1999): A Neural Network Model for Predicting the Bulk-Skin Temperature Difference at the Sea Surface, *International Journal of Remote Sensing* 20 (18), 3533-3548

Ward D.P., Murray A.T. and Phinn S.R. (2000): A Stochastically Constrained Cellular Automata Model of Urban Growth, *Computers, Environment and Urban Systems* 24, 539-558

Warner T.A. and Shank M. (1997): An Evaluation of the Potential for Fuzzy Classification of Multispectral Data Using Artificial Neural Networks, *Photogrammetric Engineering and Remote Sensing* 63 (11), 1285-1294

Webster C. (1990): Rule-Based Spatial Search, *International Journal of Geographic Information Systems* 4, 241-259

Wechsler H. and Zimmerman G.L. (1988): 2-D Invariant Object Recognition Using Distributed Associative Memory, *IEEE Transactions on Pattern Analysis and Machine Intelligence* 20, 811-821

Weigend A.S., Huberman B. and Rumelhart D.E. (1990): Predicting the Future: A Connectionist Approach, *International Journal of Neural Systems* 1, 193-209

Weigend A.S., Rumelhart D.E. and Huberman B.A. (1991a): Generalization by Weight Elimination with Application to Forecasting. In: Lippman R., Moody J. and Touretzky D. (eds.) *Advances in Neural Information Processing Systems* 3, Morgan Kaufmann Publishers, San Mateo [CA], pp. 875-882

Weigend A.S., Rumelhart D.E. and Huberman B.A. (1991b): Back-Propagation, Weight-Elimination and Time Series Prediction. In: Touretzki D.S., Elam J.L., Sejnowski T.J. and Hinton G.E. (eds.) *Connectionist Models: Proceedings of the 1996 Summer School.* Morgan Kaufmann Publishers, San Mateo [CA], pp. 105-116

Welch R.M., Sengupta S.K., Goroch A.K., Rabindra P., Rangaraj N. and Navar M.S. (1992): Polar Cloud and Surface Classification Using AVHRR Imagery: An Intercomparison of Methods, *Journal of Applied Meteorology* 31, 405-420

Werbos P.J. (1994): *The Roots of Backpropagation: From Ordered Derivatives to Neural Networks and Political Forecasting*, John Wiley, New York [NY]

White H. (1990): Connectionist Nonparametric Regression: Multilayer Feedforward Networks Can Learn Arbitrary Mappings, *Neural Networks* 3, 535-550

Whitley D. and Starkweather T. (1990a): GENITOR II: A Distributed Genetic Algorithm, *Journal of Experimental and Theoretical Artificial Intelligence* 2, 189-214

Whitley D. and Starkweather T. (1990b): Optimizing Small Neural Networks Using a Distributed Genetic Algorithm, *Proceedings of the International Joint Conference on Neural Networks* [IJCNN'90], [Volume 1], Lawrence Erlbaum Associates, Hillsdale [NJ], Washington [DC], pp. 206-209

Whitley D., Gordon S. and Mathias K. (1994): Lamarkian Evolution, the Baldwin Effect and Function Optimization. In: Davidor Y., Schwefel H.-P. and Männer R. (eds.) *Parallel Problem Solving from Nature* (PPSN) III, Springer, Berlin, Heidelberg, New York, pp. 6-15

Wilkinson G.G. (1997): Open Questions in Neurocomputing for Earth Observation. In: Kanellopoulos I., Wilkinson G.G., Roli F. and Austin J. (eds.) *Neurocomputation in Remote Sensing Data Analysis*, Springer, Berlin, Heidelberg, New York, pp. 1-3

Wilkinson G.G., Fierens F. and Kanellopoulos I. (1995): Integration of Neural and Statistical Approaches in Spatial Data Classification, *Geographical Systems* 2, 1-20

Wilkinson G.G., Kontoes C. and Murray C.N. (1993): Recognition and Inventory of Oceanic Clouds from Satellite Data Using an Artificial Neural Network Technique. In: Restelli G. and Angeletti G. (eds.) *Proceedings of the International Symposium on*

'Dimethylsulphide, Oceans, Atmosphere and Climate', Belgirate, Italy, Kluwer, Dordrecht, Boston, pp. 393-399

Wilkinson G.G., Kanellopoulos I., Mehl W. and Hill J. (1993): Land Cover Mapping Using Combined Landsat Thematic Mapper Imagery and ERS-1 Synthetic Aperture Radar Imagery. In: American Society for Photogrammetry and Remote Sensing (ed.) *Proceedings of the 12th Pecora Memorial Conference 'Land Information from Space Based Systems'*, Bethesda [MD], pp. 151-158

Wilkinson G.G., Folving S., Kanellopoulos I., McCormick N., Fullerton K. and Mégier J. (1995): Forest Mapping from Multi-Source Satellite Data Using Neural Network Classifiers: An Experiment in Portugal, *Remote Sensing Reviews* 12, 83-106

Wilson A.G. (1970): *Entropy in Urban and Regional Modelling*, Pion, London

Wilson A.G. (2000): *Complex Spatial Systems*, Prentice Hall, Harlow

Wise S., Haining R. and Ma J. (1996): Regionalisation Tools for the Exploratory Spatial Analysis of Health Data. In: Fischer M.M. and Getis A. (eds.) *Recent Developments in Spatial Analysis – Spatial Statistics, Behavioural Modelling and Computational Intelligence*, Springer, Berlin, Heidelberg, New York, pp. 83-100

Witkin A. (1983): Scale-Space Filtering, *Proceedings of the International Joint Conference on Artificial Intelligence*, Karlsruhe, pp. 1019-1021

Wolpert D.H. (1990): A Mathematical Theory of Generalization, *Complex Systems* 4 (2), 151-249

Wong Y.F. (1993a): Clustering Data by Melting, *Neural Computation* 5, 89-104

Wong Y.F. (1993b): A New Clustering Algorithm Applicable to Multispectral and Polarimetric SAR Images, *IEEE Transactions on Geoscience and Remote Sensing* 31, 634-644

Xie Y. (1996): A Generalized Model for Cellular Urban Dynamics, *Geographical Analysis* 28, 350-373

Xu Z.B., Leung Y. and He H.W. (1994): Asymmetric Bidirectional Associative Memories, *IEEE Transactions on Systems, Man, and Cybernetics* 24, 1558-1564

Yahn R.S. and Simpson J.J. (1995): Applications of Neural Networks to Cloud Segmentation, *IEEE Transactions on Geoscience and Remote Sensing* 33, 590-603

Yang G., Collins M.J. and Gong P. (1998): Multisource Data Selection for Lithologic Classification with Artificial Neural Networks, *International Journal of Remote Sensing* 19 (18), 3675-3680

Yang H., van der Meer F., Bakker W. and Tan Z.J. (1999): A Back-Propagation Neural Network for Mineralogical Mapping from AVIRIS Data, *International Journal of Remote Sensing* 20 (1), 97-110

Yao X. (1991): Evolution of Connectionist Networks. In: Dartnall T. (ed.) *Preprints of the International Symposium on AI, Reasoning & Creativity*, Griffith University, Queensland, Australia, pp. 49-52

Yao X. (1993a): An Empirical Study of Genetic Operators in Genetic Algorithms, *Microprocessing and Microprogramming* 38 (1-5), 707-714

Yao X. (1993b): Evolutionary Artificial Neural Networks, *International Journal of Neural Systems* 4 (3), 203-222

Yao X. (1993c): A Review of Evolutionary Artificial Neural Networks, *International Journal of Intelligent Systems* 8 (4), 539-567

Yao X. (1995): Evolutionary Artificial Neural Networks. In: Kent A. and Williams J.G. (eds) *Encyclopedia of Computer Science and Technology*, Volume 33, Marcel Dekker, New York [NY], pp. 137-170

Yao X. (1997): The Importance of Maintaining Behavioural Link Between Parents and Offspring, *Proceedings of the IEEE International Conference on Evolutionary Computation* [ICEC'97], Indianapolis, IEEE Press, Piscataway [NJ], pp. 629-633

Yao X. (1999): Evolving Artificial Neural Networks. *Proceedings of the IEEE'89* 87 (9), IEEE Press, Piscataway [NJ], pp. 1423-1447

Yao X. and Liu Y. (1996a): Ensemble Structure of Evolutionary Artificial Neural Networks, *Proceedings of the International Conference on Evolutionary Computation* [ICEC'96], Nagoya, Japan, IEEE Press, Piscataway [NJ], pp. 659-664

Yao X. and Liu Y. (1996b): Fast Evolutionary Programming. In: Fogel L.J., Angeline P.J. and Bäck T. (eds) *Proceedings of the Fifth Annual Conference on Evolutionary Programming*, MIT Press, Cambridge [MA], pp. 451-460

Yao X. and Liu Y. (1997): A New Evolutionary System for Evolving Artificial Neural Networks, *IEEE Transactions on Neural Networks* 8 (3), 694-713

Yao X. and Liu Y. (1998a): Making Use of Population Information in Evolutionary Artificial Neural Networks, *IEEE Transactions on Systems, Man, and Cybernetics, Part B: Cybernetics* 28 (3), 417-425

Yao X. and Liu Y. (1998b): Towards Designing Artificial Neural Networks by Evolution, *Applied Mathematics and Computation* 91 (1), 83-90

Yao X. and Shi Y. (1995): A Preliminary Study on Designing Artificial Neural Networks Using Co-evolution, *Proceedings of the IEEE International Conference on Intelligent Control and Instrumentation*, IEEE Singapore Section, Singapore, pp. 149-154

Yao X., Liu Y. and Darwen P. (1996): How to Make Best Use of Evolutionary Learning. In: Stocker R., Jelinek H. and Durnota B. (eds.) *Complex Systems: From Local Interactions to Global Phenomena*, IOS Press, Amsterdam, pp. 229-242

Yao X., Liu Y. and Lin G. (1999): Evolutionary Programming Made Faster, *IEEE Transactions on Evolutionary Computation* 3 (2), 82-102

Yoshida T. and Omatu S. (1994): Neural Network Approach to Land Cover Mapping, *IEEE Transactions on Geoscience and Remote Sensing* 32, 1103-1109

Zadeh L. (1965): Fuzzy Sets, *Information and Control* 8, 338-353

Zell A. (1994): *Simulation Neuronaler Netze*, Addison-Wesley, New York

Zhang L. and Hoshi T. (1994): A Fuzzy Neural Network Model (FNN Model) for Classification Using Landsat-TM Image Data, *Proceedings of the International Geoscience and Remote Sensing Symposium* [IGARSS'94], [Volume 3], Pasadena, IEEE Press, Piscataway [NJ], pp. 1413-1415

Zhang M. and Scofield R.A. (1994): Artificial Neural Network Technique for Estimating Heavy Convective Rainfall and Recognizing Cloud Mergers from Satellite Data, *International Journal of Remote Sensing* 15 (16), pp. 3241-3261

Zhang W.B. (1988): The Pattern Formation of an Urban System, *Geographical Analysis* 20, 75-84

Zhuang X., Engel B.A., Baumgardner M.F. and Swain P.H. (1991): Improving Classification of Crop Residues Using Digital Land Ownership Data and Landsat TM Imagery, *Photogrammetric Engineering and Remote Sensing* 57, 1487-1492

Zimmermann H.G. (1994): Neuronale Netze als Entscheidungskalkül. In: Rehkugler H. and Zimmermann H.G. (eds.) *Neuronale Netze in der Ökonomie*, Vahlen, München, pp. 1-84

Zumkeller D. (1994): Paneluntersuchungen zum Verkehrsverhalten, Forschung Straßenbau und Straßenverkehrstechnik, Volume 688, Bonn

List of Figures

List of Tables

APPENDIX

Subject Index

Author Index

List of Contributors

Michael Batty
Centre for Advanced Spatial Analysis (CASA)
University College London
1-19 Torrington Place
Gower Street
London, WC1E 6BT
United Kingdom
e-mail: mbatty@geog.ucl.ac.uk

Manfred M. Fischer
Department of Economic Geography & Geoinformatics
Vienna University of Economics and Business Administration
Rossauer Lände 23/1
A-1090 Vienna
Austria
e-mail: manfred.fischer@wu-wien.ac.at

Sucharita Gopal
Department of Geography
Boston University
675 Commonwealth Avenue
2215 Boston, Massachussetts
USA
e-mail: suchi@bu.edu

Günter Haag
Steinbeis Transfer Centre
Applied Systems Analysis
Rotwiesenstr. 22
D-70599 Stuttgart
Germany
e-mail: haag@sofo.uni-stuttgart.de

Yee Leung
Department of Geography
The Chinese University of Hongkong
Shatin, New Territories
Hongkong
e-mail: yeeleung@cuhk.edu.hk

Graeme Wilkinson
School of Computing and Information Systems
Kingston University
Penrhyn Road
Kingston Upon Thames, KT1 2EE
United Kingdom
e-mail: G.Wilkinson@Kingston.ac.uk

Xin Yao
School of Computer Science
The University of Birmingham
Edgbaston Birmingham B15 2TT
United Kingdom
e-mail: xin@cs.bham.ac.uk